Introduction to Animal Science

Introduction to Animal Science

Vera Shaffer

SYRAWOOD
PUBLISHING HOUSE
New York

Published by Syrawood Publishing House,
750 Third Avenue, 9th Floor,
New York, NY 10017, USA
www.syrawoodpublishinghouse.com

Introduction to Animal Science
Vera Shaffer

International Standard Book Number: 978-1-64740-260-0 (Hardback)

Cataloging-in-Publication Data

Introduction to animal science / Vera Shaffer.
p. cm.
Includes bibliographical references and index.
ISBN 978-1-64740-260-0
1. Animals. 2. Zoology. 3. Animal life cycles. I. Shaffer, Vera.
QL45.2 .I58 2021
590--dc23

TABLE OF CONTENTS

PREFACE

The study of the biology of animals which are controlled by humankind is known as animal science. The production and management of animals which live on farms are also dealt with under this discipline. Some of the various animals which are studied within this field of study are livestock animals such as sheep, pigs, cattle and horses along with companion animals such as dogs and cats. Animal science also makes use of other disciplines of science such as microbiology, animal behavior, genetics, physiology, reproduction and nutrition. There are various industries where animal science is applied. A few of them are livestock breeding industry, pharmaceutical industry and animal feed industry. This book provides significant information of this discipline to help develop a good understanding of animal science and related fields. Students in this field will be assisted by this book. Coherent flow of topics, student-friendly language and extensive use of examples make it an invaluable source of knowledge.

A detailed account of the significant topics covered in this book is provided below:

Chapter 1- The branch of science which is involved in the study of the biology of animals which are under human control is termed as animal science. The interest areas within this field of study include animal physiology and animal nutrition. This chapter has been carefully written to provide an easy understanding of these facets of animal science.

Chapter 2- The process of raising domesticated animals in an agricultural setting in order to produce commodities and labor is called livestock farming. Poultry farming refers to the process of raising domesticated birds for meat or eggs. The topics elaborated in this chapter will help in gaining a better perspective about poultry farming as well as the different types of livestock farming such as sheep and horse farming.

Chapter 3- The biological process through which new animals are produced by their parents is termed as animal reproduction. Animal breeding refers to the controlled propagation of domesticated animals for the purpose of improving desirable qualities. This chapter discusses in detail animal reproduction as well as the different techniques used for animal breeding.

Chapter 4- The different actions of animals such as their movement along with the underlying mental processes are studied within animal behavior. Some of the areas of interest within this field are animal learning, animal emotion and animal mimicry. This chapter discusses in detail these fields of study related to animal behavior.

Chapter 5- There are a number of diseases which affect animals such as anthrax, blackleg, blue tongue, ovine brucellosis, exudative dermatitis, swine dysentery, caprine arthritis and caseous lymphadenitis. The chapter closely examines these diseases which affect different animals to provide an extensive understanding of the subject.

I would like to make a special mention of my publisher who considered me worthy of this opportunity and also supported me throughout the process. I would also like to thank the editing team at the back-end who extended their help whenever required.

Vera Shaffer

Chapter 1

Understanding Animal Science

The branch of science which is involved in the study of the biology of animals which are under human control is termed as animal science. The interest areas within this field of study include animal physiology and animal nutrition. This chapter has been carefully written to provide an easy understanding of these facets of animal science.

Animal husbandry, also known as animal science, is the agricultural practice of breeding and raising livestock. An outgrowth of agricultural practices and techniques of farmers for centuries, the commercialization of agriculture and the advances of veterinary science during the twentieth century helped to establish a recognized scientific discipline that is taught in many universities and colleges and used all over the world.

Because of agriculture, cities, as well as trade relations between different regions and groups of people, developed, further enabling the advancement of human societies and cultures. Animal husbandry, as well as other forms of agriculture, has been an important aspect of economics throughout the centuries prior to and after the Industrial Revolution. Sustainable development of world food supplies impacts the long-term survival of human beings and all domesticated species, so care must be taken to ensure that agricultural methods remain in harmony with the environment. Breeding animals requires adequate care and concern for the impact of genetic modification, pollution, and hybridization on the species. The ethical aspects of breeding animals in captivity have also become matters of greater concern as humankind advances in awareness of its responsibility toward the environment and all living things.

Domestication

Domestication refers to the process of taming a population of animals (although it can also be used to refer to plants) or even a species as a whole. Human beings have brought these populations under their care for a wide range of reasons: To produce food or valuable commodities (such as wool, cotton, or silk), for help with various types of work, transportation, and to enjoy as pets or ornamental plants. Animals domesticated for home companionship are usually called pets while those domesticated for food or work are called livestock or farm animals.

According to evolutionary biologist Jared Diamond, animal species must meet six criteria in order to be considered for domestication:

1. Flexible diet—Creatures that are willing to consume a wide variety of food sources and can live off less cumulative food from the food pyramid (such as corn or wheat) are less expensive to keep in captivity. Most carnivores can only be fed meat, which requires the expenditure of many herbivores.
2. Reasonably fast growth rate—Fast maturity rate compared to the human life span allows

breeding intervention and makes the animal useful within an acceptable duration of caretaking. Large animals such as elephants require many years before they reach a useful size.

3. Ability to be bred in captivity—Creatures that are reluctant to breed when kept in captivity do not produce useful offspring, and instead are limited to capture in their wild state. Creatures such as the panda and cheetah are difficult to breed in captivity.
4. Pleasant disposition—Large creatures that are aggressive toward humans are dangerous to keep in captivity. The African buffalo has an unpredictable nature and is highly dangerous to humans. Although similar to domesticated pigs in many ways, American peccaries and Africa's warthogs and bush pigs are also dangerous in captivity.
5. Temperament which makes it unlikely to panic—A creature with a nervous disposition is difficult to keep in captivity as they will attempt to flee whenever they are startled. An animal such as the gazelle, which is very flighty and has a powerful leap that allows it to escape an enclosed pen is not suitable for domestication.
6. Modifiable social hierarchy—Social creatures that recognize a hierarchy of dominance can be raised to recognize a human as its pack leader. Bighorn sheep cannot be herded because they lack a dominance hierarchy, whilst antelopes and giant forest hogs are territorial when breeding and cannot be maintained in crowded enclosures in captivity.

Limits of Domestication

Hereford cattle, domesticated for beef production.

A herding instinct arguably aids in domesticating animals: Tame one and others will follow, regardless of chiefdom. Yet, despite long enthusiasm about revolutionary progress in farming, few crops and probably even fewer animals ever became domesticated.

Domesticated species, when bred for tractability, companionship, or ornamentation rather than for survival, can often fall prey to disease: Several sub-species of apples and cattle, for example, face extinction; and many dogs with very respectable pedigrees appear prone to genetic problems.

Domestication has also threatened human beings with disease: Cattle have given people various viral poxes, measles, and tuberculosis; pigs gave influenza; and horses the rhinoviruses. Humans share over sixty diseases with dogs. Many parasites also have their origins in domestic animals.

Hybridization and Genetic Engineering

Animals of domestic origin as well as feral animals sometimes produce fertile hybrids with native, wild animals which leads to genetic pollution in the naturally evolved wild gene pools, many times threatening rare species with extinction. Cases include the mallard duck, wild boar, the rock dove or pigeon, the Red Junglefowl (Gallus gallus) (ancestor of all chickens), and carp. Another example is the dingo, itself an early feral dog, which hybridized with dogs of European origin.

Genetic pollution is a serious issue: Living organisms can also be defined as pollutants, when a non-indigenous species (plant or animal) enters a habitat and modifies the existing equilibrium among the organisms of the affected ecosystem (sea, lake, and river). Non-indigenous, including transgenic species (GMOs), may bring about a particular version of pollution in the vegetal kingdom: So-called genetic pollution. This term refers to the uncontrolled diffusion of genes (or transgenes) into genomes of plants of the same type or even unrelated species where such genes are not present in nature. For example, a grass modified to resist herbicides could pollinate conventional grass many miles away, creating weeds immune to the most widely used weed-killer, with obvious consequences for crops. Genetic pollution is at the basis of the debate on the use of GMOs in agriculture.

A Genetically Modified Organism (GMO) is an organism whose genetic material has been altered using the genetic engineering techniques generally known as recombinant DNA technology. Genetic Engineering today has become another serious and alarming cause of genetic pollution because artificially created and genetically engineered animals in laboratories, which could never have evolved in nature even with conventional hybridization, can live and breed on their own and, what is even more alarming, interbreed with naturally evolved wild varieties.

Common Sub-divisions of Animal Husbandry

Historically, animal husbandry is broken up into animal subdivisions, agriculturists tending to specialize on one or two types of animals.

- Swineherd: A person who cares for hogs and pigs (older English term: Swine).
- Shepherd: Usually a person who cares for sheep. However, in previous years, it was common to have herds which were made up of both sheep and goats, in which case the term shepherd was still employed.
- Goatherd: Someone who cares primarily for goats.

- Cowherd/Cowboys: Refers to those who raise, milk, and slaughter cattle. In more modern times, the cowboys or vaqueros of North and South America ride horses and participate in cattle drives to watch over cows and bulls raised primarily for food.
- Herder: A generic term for anyone who tends to a herd of animals not listed above, such as Camels, yaks, and in Latin America, llamas and alpacas.
- Equestrian: A person who breeds and takes care of horses.
- Breeders: Generic term for anyone who specializes in the breeding of agricultural animals. More often, breeders are trained in specialized techniques, such as artificial insemination and embryo transfer, and are therefore independent hires of the usual farm staff.

Ethical Aspects of Animal Husbandry

Ethical husbandry: A wounded Norwegian reindeer is moved on a snowmobile sled.

There are contrasting views on the ethical aspects of breeding animals in captivity, with one debate being in relation to the merits of allowing animals to live in natural conditions reasonably close to those of their wild ancestors, compared to the view that considers natural pressures and stresses upon wild animals from disease, predation, and the like as vindication for captive breeding.

Some techniques of animal husbandry such as factory farming, tail docking, the Geier Hitch, and castration, have been attacked by animal welfare groups as inhumane, unsanitary, and counter-productive to the overall benefit of society.

Some of these practices also are criticized by farmers who use more traditional or organic practices. Genetic engineering is also controversial, though it does not necessarily involve suffering. People who believe in animal rights generally oppose all forms of animal husbandry.

Animal Physiology

Physiology refers to the systems and functions of the body. What are the systems in an animal's body? The major ones include the skeletal, muscular, circulatory, respiratory, nervous, urinary, endocrine, digestive, reproductive, and mammary systems.

Skeletal System

The skeletal system consists of bones and cartilage that give the body shape. It protects internal organs and provides support for the body. Bone is the major component of the skeletal system, which consists partially of calcium and phosphate. The material inside bone is bone marrow, which produces blood cells. Cartilage provides support and serves as a cushioning effect for joints.

Circulatory System

The main components of the circulatory system are the heart, arteries, capillaries, and veins. The circulatory system delivers food and oxygen to the cells of the body and eliminates waste products. It contains blood, which consists of red blood cells, white blood cells, and platelets. Red blood cells carry oxygen while white blood cells help the body fight disease.

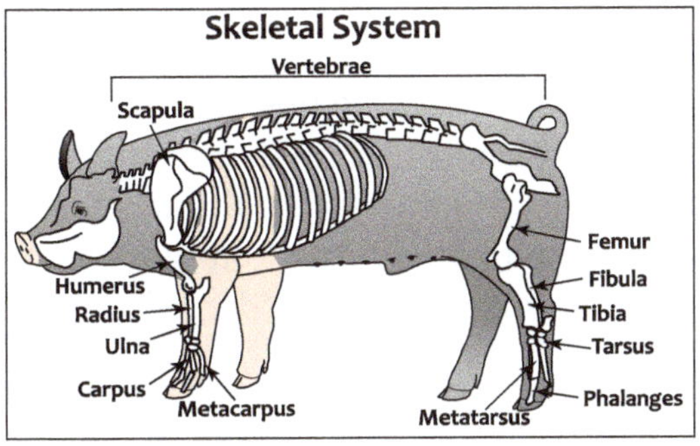

The heart, a muscular organ, pumps blood through the system. The vessels that carry blood from the heart to the body are arteries. Capillaries are small vessels that carry blood to individual cells where oxygen and nutrients are delivered, and carbon dioxide and wastes enter the blood. Veins carry blood back to the heart.

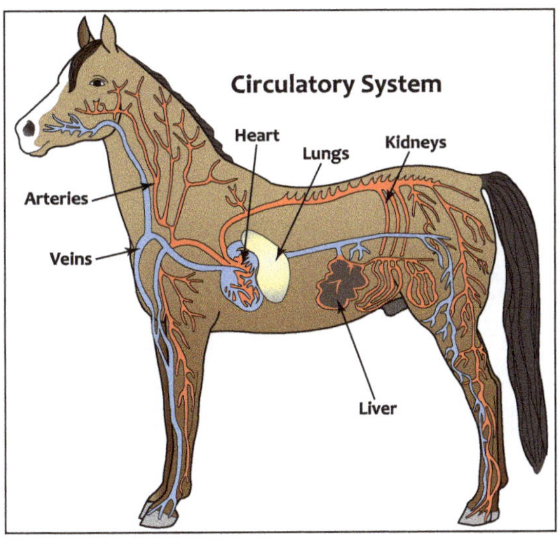

Urinary System

The urinary system, also known as the excretory system, consists of the kidneys, bladder, urethra, and ureters. The purpose of the urinary system is to remove waste products from the blood. The urinary system filters waste products and some water through the kidneys. The urine then goes through the ureters to the bladder where it is emptied through the urethra.

Muscular System

The muscular system allows movement and necessary functions such as circulation, digestion, and breathing. It is made up of muscles and connective tissue, such as tendons. The muscular system also helps provide support and shape. Muscles consist largely of protein and are the lean meat of an animal that is used as food for human consumption. Muscles can be voluntary or involuntary. Voluntary muscles are those that can be controlled. When an animal walks or chews it is using voluntary muscles. Involuntary muscles are automatic and operate without any thought on the part of an animal. Muscles that operate the heart, intestines, and lungs are involuntary.

Respiratory System

The major parts of the respiratory system are the nostrils, nasal cavity, pharynx, larynx, trachea, and lungs. When using the respiratory system, animals breathe in needed oxygen and breathe out carbon dioxide. Air first enters the respiratory system though the nostrils or mouth. Then the air passes through the nasal cavity to the pharynx. Finally, air goes through the trachea and enters the lungs. In the lungs, air is exchanged for carbon dioxide. When animals breathe out, this carbon dioxide is removed from the body.

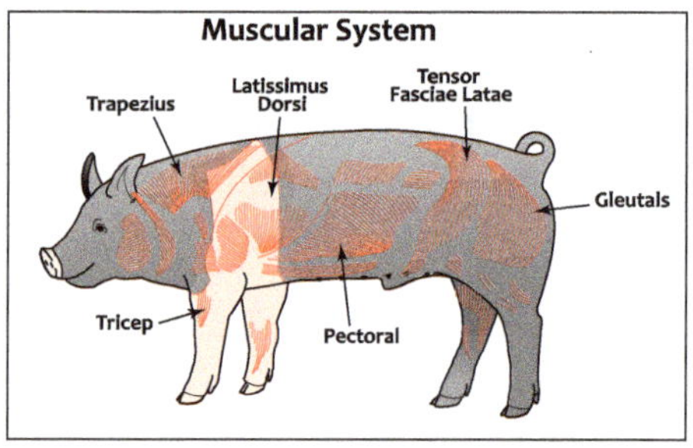

Nervous System

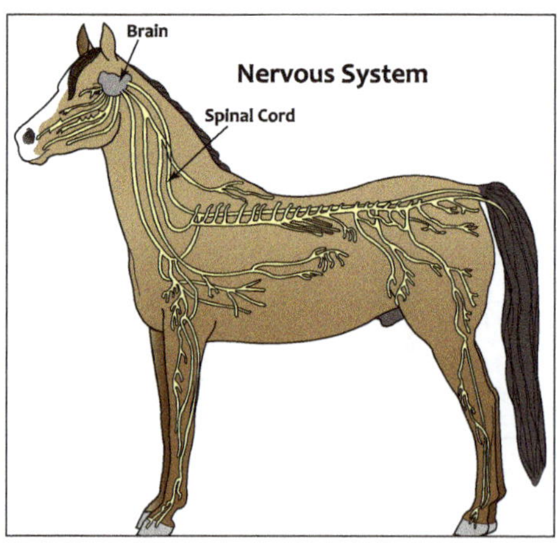

The nervous system allows the body to communicate by sending, receiving, and interpreting signals. The nervous system is composed of the brain, spinal cord, and neurons. The nervous system can be divided into two parts: The central nervous system and the peripheral nervous system. The central nervous system includes the brain and spinal cord. It is responsible for receiving information from nerves that pass through the spinal cord and others that control the senses of hearing, sight, smell, touch and taste.

The peripheral nervous system connects the central nervous system to the rest of the body. The peripheral nervous system is the system that is responsible for voluntary muscle control. The autonomic nervous system is a part of the peripheral nervous system and controls automatic body activities, such as the heart muscle.

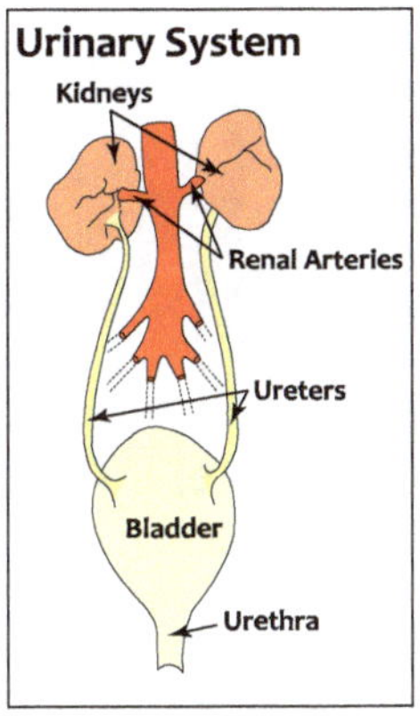

Endocrine System

The endocrine system consists of endocrine glands that release hormones into the body. Hormones are chemicals necessary for the regulation of many bodily functions. For example, hormones regulate growth and reproduction.

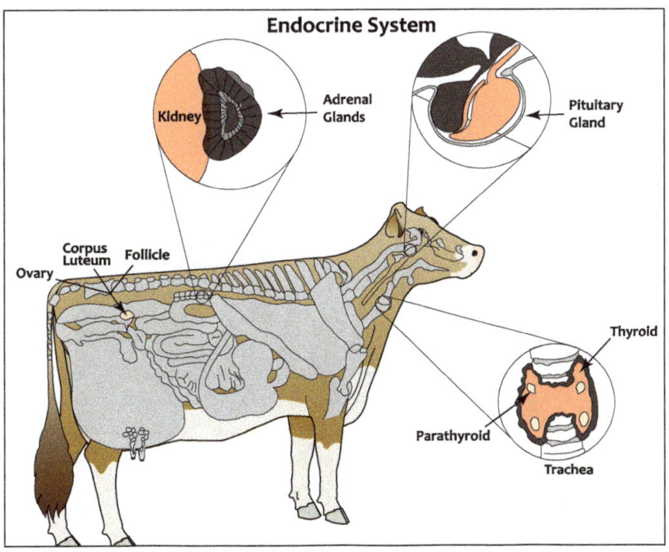

Digestive System

The digestive system breaks food down so that it can be used by the body. Digestive systems can vary depending on the species of animal. Some have very simple stomachs and others are more complex.

Reproductive and Mammary Systems

The reproductive system is what enables reproduction to occur. The male and female reproductive systems differ in animals. The female reproductive system provides an environment for gestation. In female animals, the ovaries produce eggs, which are usually released on a regular cycle through ovulation. Eggs can become fertilized when mating occurs during a certain period of the ovulation cycle.

Nursing mothers require a larger amount of nutrients in order to feed their young.

The male reproductive system provides the sperm necessary for a female to become pregnant through fertilization. Animals must reach a certain age before they are able to reproduce safely.

Milk is produced through the mammary system of females. The process of producing milk is known as lactation and occurs after giving birth and stops when the offspring are weaned. In many mammals, the mammary system initially produces colostrum, a fluid that contains antibodies for the new born. Also known as the first milk, colostrum aids young animals in fighting off infections due to the introduced antibodies.

Animal Nutrition

Some animals need just enough nutrients to maintain current body weight.

Nutrition is important for a variety of reasons. Animals need the proper nutrition for growth and maintenance, and to provide energy for work and vital functions. Maintenance is the nutrition required for an animal to maintain its current weight. Energy is the ability of the body to perform functions. Proper nutrition is also needed to maintain body temperature, produce milk, reproduce, and develop proper bone structures. Without proper nutrition, animals can develop health problems, which could result in treatment costs or even fatality. Good nutrition is essential for all of the systems of an animal to function and work together properly.

Animals that do not receive the proper nutrition are more likely to develop health and reproductive problems, and be less productive and marketable. Nutrition is important to have healthy animals, and, in turn, healthy animals help ensure profitability in agricultural operations.

Producers must understand not only what nutrition an animal needs but also how to supply that needed nutrition. Producers must know what ration is appropriate for a specific animal and how to provide a balanced ration. A ration is the feed an animal receives over a 24-hour period. A variety of feedstuffs, or basic ingredients, are used in rations, and producers must choose those that best suit their needs. A ration with all the nutrients an animal needs is a balanced ration.

Animal Feed for Nutrition

Animal Feed refers to the food grown or developed for livestock and poultry. Modern feeds are produced by carefully selecting and blending ingredients to provide highly nutritional diets that both maintain the health of the animals and increase the quality of such end products as meat, milk, or eggs. On-going improvements in animal diets have resulted from research, experimentation, and chemical analysis by agricultural scientists.

Animals in general require the same nutrients as humans. Some feeds, such as pasture grasses, hay and silage crops, and certain cereal grains, are grown specifically for animals. Other feeds, such as sugar beet pulp, brewers' grains, and pineapple bran, are by-products that remain after a food crop has been processed for human use. Surplus food crops, such as wheat, other cereals, fruits, vegetables, and roots, may also be fed to animals.

History does not record when dried roughage or other stored feeds were first given to animals. Most early records refer to nomadic peoples who, with their herds and flocks, followed the natural feed supplies. When animals were domesticated and used for work in crop production, some of the residues were doubtless fed to them.

The first scientific effort to evaluate feeds for animals on a comparative basis was probably made in 1809 by the German agriculturist Albrecht von Thaer, who developed "hay values" as measures of the nutritive value of feeds. Tables of the value of feeds and of the requirements of animals in Germany followed and were later used in other countries.

Preservation of green forages such as beet leaves and corn (maize) plants by packing them in pits in the earth has long been practiced in northern Europe. The idea of making silage as a means of preserving and utilizing more of the corn plant was gradually developed in Europe and was taken from France to the United States in the 1870s. When the mature, dried corn plant was fed to cattle in the winter, much of the coarse stem was wasted, but when it was chopped and ensiled (made into silage), everything was eaten. During the 20th century, concrete bunker silos for storage of silage became a common sight in many rural areas worldwide.

Basic Nutrients and Additives

The basic nutrients that animals require for maintenance, growth, reproduction, and good health include carbohydrates, protein, fat, minerals, vitamins, and water. The energy needed for growth and activity is derived primarily from carbohydrates and fats. Protein will also supply energy, particularly if carbohydrate and fat intake is inadequate or if protein intake exceeds the needs of the body.

Animals need a source of energy to sustain life processes within the body and for muscular activity. When the energy intake of an animal exceeds its requirements, the surplus is stored as body fat, which can be utilized later as a source of energy if less food becomes available.

Proteins

For immature animals, protein is also needed for growth of the muscles and other parts of the body. Since milk, eggs, and wool contain much protein, additional amounts are needed in the feed of animals producing these. All animals require a small amount of protein for maintenance—i.e., the daily repair of muscles, internal organs, and other body tissues.

Proteins are composed of more than 20 different amino acids, which are liberated during digestion. Animals with a simple single stomach (monogastric), including humans, monkeys, swine, poultry, rabbits, and mink, require correct amounts of the following 10 essential amino acids daily: arginine, histidine, isoleucine, leucine, lysine, methionine, phenylalanine, threonine, tryptophan, and valine. In addition to these, poultry need glycine and glutamic acid for growth. Cystine can replace up to half of the methionine requirement, and tyrosine can replace up to half of the phenylalanine requirement. High-quality protein, such as that supplied by eggs, milk, fish meal, meat by-products, and soybean meal, contains high concentrations of the essential amino acids in the proper balance for their full utilization. Poor-quality protein, such as that in most grains, including corn, barley, and sorghum, contains too little of one or more essential amino acids. Feeds having poor-quality proteins are useful when blended with other feeds that restore the balance in essential amino acids.

A protein source's amino acid profile is of secondary importance to ruminants, such as cattle, sheep, goats, and the other animals that have four stomachs, because the bacteria that aid in the digestion of food in the rumen (first stomach) use simple nitrogen compounds to build proteins in their cells. Further on in the digestive tract, the animals digest the bacteria. By this indirect means, ruminants produce high-quality protein from a food that might originally have contained poor protein or from urea (a nitrogen compound). Very young ruminants, such as calves, lambs, and kids, however, need good-quality protein until the rumen develops sufficiently for this bacterial process to become established.

Carbohydrates and Fats

Most animals get energy from carbohydrates and fats, which are oxidized in the body. These yield heat, which maintains body temperature, furnishes energy for growth and muscle activity, and sustains vital functions. Animals need much more energy (and more total feed) for growth, work, or milk production than for simple maintenance.

Simple carbohydrates such as sugars and starches are readily digested by all animals. The complex carbohydrates (cellulose, hemicelluloses) that make up the fibrous stems of plants are broken down by bacterial and protozoal action in the rumen of cattle and sheep or in the cecum of rabbits and horses. Such complex carbohydrates cannot be digested by humans or, to any appreciable extent, by dogs, cats, birds, or laboratory animals. Thus, ruminants and some herbivorous animals obtain much more of the energy-giving nutrients from the carbohydrates of plants than do monogastric carnivores and omnivores, for which fibrous materials have little or no energy value.

Fat in feeds has a high nutritive value because it is easily digested and because it supplies about two and one-quarter times as much energy as an equal weight of starch or sugar. While fat has a high nutritive value, it can be replaced by an equivalent amount of digestible carbohydrates in the feed, except for small amounts of essential fatty acids. Very small amounts of the unsaturated fatty acid linoleic, contained in some fats, are necessary for growth and health. Animal feeds typically supply ample amounts of this acid unless it has been removed by processing.

Minerals

Minerals essential for animal life include common salt (sodium chloride), calcium, phosphorus, sulfur, potassium, magnesium, manganese, iron, copper, cobalt, iodine, zinc, molybdenum, and selenium. The last six of these can be toxic to animals if excessive amounts are eaten.

All farm animals generally need more common salt than is contained in their feeds, and they are supplied with it regularly. Of the other essential minerals, phosphorus and calcium are most apt to be lacking, because they are heavily drawn upon to produce bones, milk, and eggshells. Good sources of calcium and phosphorus are bonemeal, dicalcium phosphate, and defluorinated phosphates. Eggshells are nearly pure calcium carbonate. Calcium may readily be supplied by ground limestone, ground seashells, or marl, which are all high in calcium.

Small amounts of iodine are needed by animals for the formation of thyroxine, a compound containing iodine, secreted by the thyroid gland. A serious deficiency of iodine may cause goitre, a disease in which the thyroid gland enlarges greatly. In certain regions, goitre has caused heavy losses of new born pigs, lambs, kids, calves, and foals. Iodine deficiencies can be prevented by supplying iodized salt to the mother before the young are born. Almost all commercial sources of salt for animals contain iodine as a routine additive.

In some areas, soil and forage are deficient in copper and cobalt, which are needed along with iron for the formation of hemoglobin, the oxygen-carrying pigment of the red blood cells. In these areas, farm animals may suffer from anemia unless the deficiency is corrected by means of a suitable mineral supplement. Ruminants are usually fed cobalt in the diet so they can then synthesize vitamin B_{12}. Monogastric animals, such as pigs, require a direct source of vitamin B_{12} in their diet.

Iron, used in hemoglobin formation, is amply supplied in most animal feeds, except milk. The only practical problem with iron deficiency occurs in young suckling pigs before they start to consume other feeds in addition to milk. They require an iron injection or access to fresh soil to meet their iron needs.

Manganese is essential for animals, and the usual diets for all farm animals supply sufficient quantities. A lack of manganese may cause the nutritional disease of chicks and young turkeys called slipped tendon (perosis) and also may cause failure of eggs to hatch. Normal diets for swine are often deficient in zinc, especially in the presence of excess calcium. Fortifying feed by adding 100 parts per million of zinc, as zinc sulfate or zinc carbonate, prevents zinc deficiency symptoms, which include retarded growth rate and severe scaliness and cracking of the skin (parakeratosis). While trace amounts of selenium are necessary for normal health, excessive amounts, which can be found in forages and grains in some regions, are toxic and may cause death. To furnish both calcium and phosphorus, grazing livestock may be allowed free access to such a mixture as 60 percent dicalcium phosphate and 40 percent common salt. Trace mineralized salt is used when copper or cobalt may be deficient. Livestock usually are given access to common salt separately, so they will not be forced to eat more of the other minerals than they require to get the amount of salt they need. Swine diets usually contain prescribed levels of calcium, phosphorus, salt, and essential trace minerals that may be deficient in the grains they are fed.

Vitamins

Known vitamins include the fat-soluble vitamins A, D, E, and K, and the water-soluble B group of thiamin, riboflavin, niacin, pantothenic acid, choline, biotin, folic acid, and vitamins B_6 and B_{12} and vitamin C.

Vitamin A, the one most apt to be lacking in livestock feeds, is required for growth, reproduction, milk production, and the maintenance of normal resistance to respiratory infections. All green-growing crops are rich in carotene, which animals can convert into vitamin A. Vitamin A supplement is added to animal diets to ensure a supply when livestock are not fed green forages and are not on good pasture.

Vitamin D enables animals to use calcium and phosphorus; a deficiency causes rickets in young growing animals. The ultraviolet rays of sunlight produce vitamin D from the provitamin in the skin. Field curing of hay develops vitamin D through the action of the sunlight on ergosterol in the hay crops. Certain fish oils are very rich in vitamin D. Livestock that are outdoors in the sunlight much of the time have a plentiful supply of vitamin D. Under winter conditions in cold regions, cattle, sheep, and horses ordinarily get ample amounts from the hay they are fed; pigs, poultry, and laboratory animals that are raised indoors will be deficient unless a supplement is added.

The vitamin B group is not important in the feeding of cattle, sheep, and other ruminants, because the bacteria in their rumen synthesize these vitamins. Very young calves, however, and poultry, swine, and other monogastric animals require the B vitamins in their diets. Of these, riboflavin, niacin, pantothenic acid, and vitamin B_{12} are most likely to be deficient in ordinary feeds; special supplements are needed by pigs, poultry, and laboratory animals. Choline may also be deficient in poultry feeds.

Vitamin E is necessary for normal hatching of eggs. It plays a role along with selenium in preventing muscle stiffness and paralysis (dystrophy) in lambs, calves, and chicks under certain conditions. Vitamin C, which prevents scurvy in humans and guinea pigs, can be synthesized in the bodies of most other animals and need not be supplied in their food. Vitamin K is synthesized by bacteria in the intestinal tract and can be absorbed, and, if livestock can ingest feces, a dietary supply is usually not important. Today many animals are raised without fecal contact, though, so vitamin K is often added to their diets as a safety factor.

Antibiotics and other Growth Stimulants

Antibiotics have been used in livestock diets since the early 1950s. They and other growth stimulants are non-nutritive substances added to animal feeds to treat diseases, to improve the efficiency of feed utilization and feed acceptance, or to improve the health or metabolism of the animal in some way. The use of antibiotics can be broadly divided into two categories, therapeutic and sub-therapeutic, in which the distinction purely depends on the amount added to the feed. In therapeutic use, enough antibiotics are used to control bacterial infections within an individual or animal population; in sub-therapeutic use, antibiotics are given in relatively low doses to enhance the performance (typically growth and feed efficiency) of the animals. The addition of sub-therapeutic antibiotics to the diets of young pigs improves growth performance by 10 to 15 percent or more. Because the sub-therapeutic use of antibiotics does not completely eradicate bacteria populations, over time the practice leads to antibiotic-resistant bacteria, which forces the search for new antibiotics for both livestock and humans. Some countries in the European Union have banned the sub-therapeutic use of antibiotics, and there is growing pressure in the United States to ban the sub-therapeutic use of penicillin and tetracyclines, the most important antibiotics for human use. The most commonly used antibiotics in feed are chlortetracycline, oxytetracycline, bacitracin, penicillin, and tylosin.

Other growth enhancers added to the feed do not have a disease-reducing affect but rather change the animal's metabolism. Most are related to hormones produced by the animals. These include lasalocid for cattle and sheep, melengestrol acetate and monensin for cattle, and ractopamine for swine. Several ear implants are approved for delivering hormones or other drugs to feedlot cattle in the United States. For example, these products typically increase daily gain by 10 to 15 percent and feed efficiency by 5 to 10 percent. Other countries, particularly those in the European Union, severely restrict or prohibit the use of implants in meat-producing animals because of opposition from some consumer groups.

Composition and Valuation of Feeds

The usual chemical analyses of feeds provide information on the amount of dry matter, protein (with its amino acid composition), fat, fibre, minerals, and vitamins contained in the feed. Various energy values (digestible, metabolizable, and net) of the feed, which depend on the species of animal being fed, are included in complete tables of feed composition.

Determination

Digestion and balance experiments measure the degree to which the various components of a feed are absorbed and retained by the animal body. Proteins are composed of varying quantities of amino acids, which are ultimately used by the animal to develop muscle and other body tissues. Microbes in the rumens of cattle and sheep can synthesize amino acids from the various nitrogen sources in their feed. In contrast, monogastrics such as pigs and poultry must have the proper amounts of the essential amino acids provided in their diet. Thus, ruminants typically have simple protein requirements, while monogastrics have amino acid requirements.

Protein and amino acid requirements are expressed as the amounts of digestible protein or amino acids needed for growth or other body functions, either as a percentage of the diet or as the total grams or units required per day. The amounts of energy needed are measured as digestible energy (DE), metabolizable energy (ME), net energy (NE), or total digestible nutrients (TDN). These values differ with species. The gross energy (GE) value of a feed is the amount of heat liberated when it is burned in a bomb calorimeter. The drawback of using this value is that a substance such as wood and corn may have a similar GE but vastly different nutritional values because some or all of it passes through the body without being digested. Furthermore, some of what is digested gets excreted in the urine as urea. Still more energy is lost from the heat of digestion and as gas produced by bacteria in the digestive tract. This loss is appreciably greater in ruminants than in pigs, chickens, and other monogastric animals.

The work of eating, digesting, and metabolizing food may also be subtracted from the food energy, resulting in the NE value, or useful energy value of a feed that can be used for production (growth, reproduction, or milk) or for maintenance (basal metabolism, activity, keeping warm). The TDN value of a feed represents the sum of the digestible protein, digestible ether extract (fat) times 2.25, digestible nitrogen-free extract (carbohydrate), and digestible crude fibre. Its use as a measurement factor has declined in recent years because it considers neither the fermentation and heat losses during digestion and metabolism nor the fact that energy is utilized more efficiently for maintenance or milk production than for growth and fattening.

Optimization of Nutrient-cost Ratio

Feed costs vary widely from season to season: it is often possible for producers to realize substantial savings through wise selection of the feed ingredients used to formulate complete diets. It is much easier for large commercial feed companies with widespread operations to take advantage of regional variations in feed prices than it is for individual relatively small-scale livestock producers who must rely on locally produced feeds.

Least-cost formulation of feed mixtures makes it possible to use computers to select the correct amounts of competitively priced feed ingredients that will combine to fully satisfy the nutrient requirements of a specific type of animal at a particular stage of development. When used by a qualified nutritionist, computer programs can successfully formulate diets that yield maximum production at the lowest possible cost.

Basic Types of Feeds

Animal feeds are classified as follows: concentrates, high in energy value, including fat, cereal grains and their by-products (barley, corn, oats, rye, wheat), high-protein oil meals or cakes (soybean, canola, cottonseed, peanut [groundnut]), and by-products from processing of sugar beets, sugarcane, animals, and fish, and roughages, including pasture grasses, hays, silage, root crops, straw, and stover (cornstalks).

Concentrate Foods

Cereal Grains and their By-products

In the agricultural practices of North America and northern Europe, barley, corn, oats, rye, and sorghums are grown almost entirely as animal feed, although small quantities are processed for human consumption as well. These grains are fed whole or ground, either singly or mixed with high-protein oil meals or other by-products, minerals, and vitamins to form a complete feed for pigs and poultry or an adequate dietary supplement for ruminants and horses.

The production of grains is seasonal because of temperature or moisture conditions or a combination of both. It is necessary to produce a full year's supply during the limited growing season. The grain is dried to 14 percent or less moisture to prevent sprouting or molding; the grain is then stored in containers or buildings where insects and rodents cannot destroy it. It is generally desirable to store more than a year's supply of the grains to be used as feed, because crop failures sometimes occur.

High-protein Meals

Vegetable seeds produced primarily as a source of oil for human food and industrial uses include soybeans, peanuts (groundnuts), flaxseed (linseed), canola, cottonseed, coconuts, oil palm, and sunflower seeds. After these seeds are processed to remove the oil, the residues, which may contain from 5 percent to less than 1 percent of fat and 20 to 50 percent of protein, are marketed as animal feeds. Cottonseed and peanuts have woody hulls or shells, which are generally removed before processing—if the hulls or shells are left intact, the resulting by-product is higher in fibre and appreciably lower in protein and energy value. The protein quality of these meals for monogastrics varies greatly depending on the levels and availability of the amino acids present. Ruminants in general require only protein or nitrogen sources for the rumen microbes to synthesize amino acids.

These high-protein feeds supplement inexpensive roughages, cereal grains, and other low-protein feeds in order to furnish the protein and amino acids needed for efficient growth or production. The supplement chosen for a particular diet depends largely on the cost and availability of supply.

By-products of Sugar Beets and Sugarcane

From the sugar beet industry come beet tops, which are used on the farm either fresh or ensiled, and dried beet pulp and beet molasses, which are produced in sugar factories. Cane molasses is a residue from cane sugar manufacture. These are all palatable, high-quality sources of carbohydrates. Sugarcane bagasse (stalk residue) is fibrous, hard to digest, and of very low feed value. In Europe, beets and some other roots are grown as animal feed. Citrus molasses and dried citrus pulp, which are generally available at low cost as by-products of the citrus juice industry, are often used as high-quality feeds for cattle and sheep.

Other By-product Feeds

Large quantities of animal feed are by-products or residues from commercial processing of cereal grains for human consumption. The largest group of these by-product feeds comes from the milling of wheat, including wheat bran, wheat middling, wheat germ meal, and wheat mill feed. In some areas, bakery wastes, such as stale and leftover bread, rolls, and various pastry products, are ground and used as filler or feed for pets and farm animals. Rice bran and rice hulls are obtained in similar fashion from the mills that polish rice for human food. Corn gluten feed, corn gluten meal, and hominy feed are produced as by-products from the manufacture of starch for industrial and food uses.

Brewers' grains, corn distillers' grains and solubles, and brewer's yeast are useful animal feeds and are collected from the dried residues of the fermentation industries that produce beer and distilled spirits. Waste products from pineapple-canning plants include pineapple bran or pulp and the ensiled leaves from the plant. By-products from the abattoirs and meatpacking plants that process animals into meat include such feeds as meat and bonemeal, tankage (animal residue left after rendering fat in a slaughterhouse), meat scraps, blood meal, poultry waste, and feather meal. Various types and qualities of fish meals are produced by fish-processing plants. These animal by-products typically contain 50 percent or more high-quality protein and the mineral elements calcium and phosphorus. Steamed bonemeal is particularly high in these important minerals. Dried skim milk, dried whey, and dried buttermilk are feed by-products from the dairy industry.

Roughages

Pasture

Pasture grasses and legumes, both native and cultivated, are the most important single source of feed for ruminants such as cattle, horses, sheep, and goats. During the growing season they furnish most of the feed for these animals at a cost lower than for feeds that need to be harvested, processed, and transported. Hundreds of different grasses, legumes, bushes, and trees are acceptable as feeds for grazing animals. The nutritive value of the cultivated varieties has been studied, but information is incomplete for many of those that occur naturally.

Hay

Hay is produced by drying grasses or legumes when they approach the stage of maximum plant growth and before the seed develops. This stage has been shown to give maximum yields of digestible protein and carbohydrates per unit of land area. The moisture content is typically reduced below 18 percent in order to prevent molding, heating, and spoilage during storage. Legume hays, such as alfalfa and clovers, are high in protein, while the grasses (such as timothy and Sudan grass) are lower in protein and vary considerably depending on their stage of maturity and the amount of nitrogen fertilization applied to them. Stored hay is fed to animals when sufficient fresh pasture grass is not available.

Bales of hay.

Silage

Silage is made by packing immature plants in an airtight storage container and allowing fermentation to develop acetic and lactic acids, which preserve the moist feed. Storage may be in upright tower silos or in trenches in the ground. The initial moisture concentration of the forage should be between 50 and 70 percent, depending on the type of silage. Lower moisture levels can cause difficulty in obtaining sufficient packing to exclude air and may result in molding or other spoilage. Too high a moisture content causes nutrient losses by seepage and results in the production of excessively acidic, unpalatable silage. Ensiled forage can be stored for a longer period of time with lower loss of nutrients than dry hay. The nutritive value of silage depends on the type of forage ensiled and how successfully it has been cured. Corn, sorghums, grasses, and sometimes leguminous forages are used in making silage.

Making silage for cattle on a farm.

Root Crops

Root crops are used less extensively as animal feed than was true in the past, for economic reasons. Beets (mangels), rutabagas, cassava, turnips, and sometimes surplus potatoes are used as feed. Compared with other feeds, root crops are low in dry-matter content and protein; they mostly provide energy.

Straw and Hulls

Quantities of straw remaining after the harvesting of wheat, oats, barley, and rice crops are used as feed for cattle and other ruminants. The straws are low in protein and very high in fibre; digestibility is low. Straw is useful in maintaining mature animals when other feeds are in short supply, but it is too low in nutrition to be a satisfactory feed for extended periods unless supplemented with other feeds that supply the protein, digestible energy, and minerals needed for growth and production. Treatment of straw with alkali markedly increases the digestibility of the cellulose, augmenting its value as a source of energy for animals.

Corncobs, cornstalks, cottonseed hulls, and rice hulls can also be used as sources of fibre in ruminant diets. Rice hulls are lower in value, while the others are similar to straw.

Chapter 2

Livestock and Poultry Farming

The process of raising domesticated animals in an agricultural setting in order to produce commodities and labor is called livestock farming. Poultry farming refers to the process of raising domesticated birds for meat or eggs. The topics elaborated in this chapter will help in gaining a better perspective about poultry farming as well as the different types of livestock farming such as sheep and horse farming.

Livestock Farming

Livestock (singular or plural) is any domesticated mammal intentionally reared in an agricultural setting for the purposes of profit or subsistence, whether for food, fiber, dairy, draft, breeding, sport purposes, or other product or labor. As such, livestock includes animals such as cattle, horses, sheep, and fur-bearing animals, but does not include farmed birds (turkeys, chickens, pigeons, geese), fish, shellfish, amphibians (frogs), and reptiles. It also does not include animals kept as pets. However, the term is not rigorously applied in all contexts and inclusion of poultry as livestock is not uncommon. And in many cases ratites (emus, ostriches, rheas) are considered as livestock when raised in an agricultural setting, even when poultry are not counted (IRS).

Raising animals (animal husbandry) is an important component of modern agriculture. It has been practiced in many societies, since the transition to farming from hunter-gather lifestyles. The ability to raise livestock has enabled the development of human societies and cultures, fostered commerce and international trade, and provided a steady source of food, labor, and other products for people. However, in recent years there has been a greater awareness of the need to address environmental concerns from the impact of livestock and their rearing as well as ethical concerns regarding the treatment of livestock.

Domestic sheep and a cow pastured together.

Livestock farming is the raising of animals for use or for pleasure. An efficient and prosperous animal agriculture historically has been the mark of a strong, well-developed nation. Such agriculture permits a nation to store large quantities of grains and other foodstuffs in concentrated form to be utilized to raise animals for human consumption during such emergencies as war or natural calamity. Furthermore, meat has long been known for its high nutritive value, producing stronger, healthier people.

Ruminant (cud-chewing) animals such as cattle, sheep, and goats convert large quantities of pasture forage, harvested roughage, or by-product feeds, as well as non-protein nitrogen such as urea, into meat, milk, and wool. Ruminants are therefore extremely important; more than 60 percent of the world's farmland is in meadows and pasture. Poultry also convert feed efficiently into protein; chickens, especially, are unexcelled in meat and egg production. Milk is one of the most complete and oldest known animal foods. Cows were milked as early as 9000 BCE. Hippocrates, the Greek physician, recommended milk as a medicine in the 5th century BCE. Sanskrit writings from ancient India refer to milk as one of the most essential human foods.

Cattle Farming

Cattle refer to any domesticated bovine farm animals that are raised for their meat, milk, or hides or for draft purposes. The animals most often included under the term are the Western or European domesticated cattle as well as the Indian and African domesticated cattle. However, certain other bovids such as the Asian water buffalo, the Tibetan yak, the gayal and banteng of Southeast Asia, and the plains bison of North America have also been domesticated or semi-domesticated and are sometimes considered to be cattle.

Ankole-Watusi cattle.

In the terminology used to describe the sex and age of cattle, the male is first a bull calf and if left intact becomes a bull; if castrated he becomes a steer and in about two or three years grows to an ox. The female is first a heifer calf, growing into a heifer and becoming a cow. Depending on the breed, mature bulls weigh 450–1,800 kg (1,000–4,000 pounds) and cows 360–1,100 kg (800–2,400 pounds). Males retained for beef production are usually castrated to make them more docile on the range or in feedlots; with males intended for use as working oxen or bullocks, castration is practiced to make them more tractable at work. The use of cattle as commodities has been a point of philosophical contention throughout history, particularly regarding the raising of animals for food. Such issues are compounded by modern concerns about the ethics of industrial factory farming and the contribution of commercial meat production to global warming.

Hereford cow and calf.

Murray Grey.

All modern domestic cattle are believed to belong to the species Bos taurus (European breeds such as Shorthorn and Jersey) or Bos indicus (Zebu breeds such as Brahman) or to be crosses of these two (such as Santa Gertrudis). Breeds as they are known today did not always exist, and many are of recent origin. The definition of a breed is difficult and inexplicit, although the term is commonly used and, in practice, well understood. It may be used generally to connote animals that have been selectively bred for a long time so as to possess distinctive identity in colour, size, conformation, and function, and these or other distinguishing characteristics are perpetuated in their progeny. Breeds have been established by generations of breeders aiming at the attainment and preservation of a particular type with its identifying characteristics. This is accomplished by working on the principle of "like begets like." It is only in relatively recent times that the science of genetics, and particularly population genetics, has contributed to breeding.

Cattle farming involve rearing and management of two types of animals- one group for food requirements like milk and another for labor purposes like plowing, irrigation, etc. Animals which provide milk are called milch/dairy animals. For example, goats, buffalo, cows, etc. Animals which are used for labor are called draught animals.

Since dairy animals are cared and bred for milk, we need to improve the milk production to meet the requirements. The period after the birth of a calf, when a cow starts to produce milk, is called lactation period. We can enhance milk production by increasing this lactation period. But along with milk production, quality must also meet. Dairy farm management is the management of the milch animals with the goal of enhancing the quantity and quality of the milk produced. For this reason, high yielding and disease resistant breeds are developed.

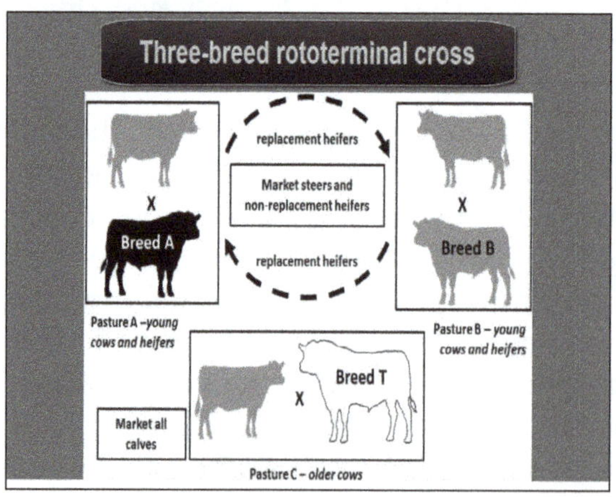

Cross-breeding in Cattle Farming.

For example, the foreign breeds like Jersey, Holstein-Friesian, Brown Swiss, have long lactation period while local breeds like Red Sindhi, Sahiwal, and Gir are known for their disease-resistant trait. The breeding of these two varieties helped us to enhance the quantity and quality of the milk produced.

Farm Management

Cattle farming are not all about milk or meat production. It has some responsibilities to do. To meet the human requirements, we need to take good care and need proper management of livestock.

Shelter

In cattle farming, animals are maintained in a strictly hygienic manner with proper housing. For the maintenance, we need to follow some routines:

- Animals and their sheds need to be cleaned at a regular interval.
- Animals should be brushed regularly to remove the dirt and bugs in their body.
- The shed should be well-ventilated and roofed so that animals are protected from rain, heat, and cold.
- A proper drainage system should be there to remove animal waste.

Food

Food requirements of animals are also a part of cattle farming. To maintain high yielding and disease-resistant breeds, they must be provided with an adequate water supply and nutrient-rich fodder regularly according to their needs. In cattle farming, two factors are considered regarding the food of cattle. The food that is provided must keep the animal health as well as it should meet the farming requirement. Hence, the animal feed includes roughage (high fiber content) and concentrates (high proteins and nutrient content). In addition to this, supplements containing micronutrients are also provided to animals. An adequate proportion of these rations promote healthy and high output animals.

Disease Management

The third responsibility of cattle farming management is to maintain disease-free breeds. Animals are not an exception to disease. They also suffer from numerous diseases. This may affect the health as well as productivity of animals; even cause their death. Parasites, bacteria, and viruses are the major villains here. These microbes infect the cattle externally as well as internally. Vaccination is the one solution for the protection against bacterial and viral infections.

Pig Farming

Pigs are even-toed ungulate mammals (Order Artiodactyla) of the genus Sus, within the Suidae family. They are native to Eurasia and northern Africa. Pigs, in the wild, live mostly in forests and partly wooded areas and play an important role in nature. The domestic pig, Sus scrofa domesticus, was one of the first animals domesticated by humans and is still today one of the most important domestic animals.

The treatment of pigs in today's intensive agriculture raises animal welfare concerns regarding this highly intelligent animal. Likewise, although wild pigs are an integral part of ecosystems, providing a number of ecological benefits, the development of feral populations of domestic pigs has resulted in environmental concerns, particularly in regions where pigs are not native.

Pigs are one of the most widespread and successful genera of large mammals. They are found wild over most of Eurasia from tropical jungles to northern forests.

There are 10 species of living pigs. The Eurasian wild boar, Sus scrofa, has the largest range of any wild ungulate and has about 25 subspecies. The smallest pig, the Pygmy hog of India, Sus salvanius, is about 20 to 30 cm (8 to 12 in) high at the shoulder and weighs 7 to 12 kg (15 to 26 lbs); the Eurasian wild boar can be from 55 to 110 cm (1.8 to 3.6 ft) high at the shoulder and weigh from 44 to 320 kg (91 to 711 lb).

Unlike most other hoofed mammals, pigs do not have multi-chambered ruminating stomachs and can not survive on leaves and grasses alone. Instead they are omnivores, which mean that they consume both plants and animals. They eat a wide variety of foods, including acorns and other seeds, green vegetation, roots, tubers, fungi, fruit, carrion, eggs, insects, and small animals—snakes being a favorite. Occasionally, in times of shortage, a mother pig may eat her own young.

Painting of wild boars.

A typical pig has a large head with a long snout, which is strengthened by a special bone called the prenasal bone and by a disk of cartilage in the tip. The snout is used to dig into the soil to find food and is a very sensitive sense organ. Pigs have a full set of 44 teeth. The canine teeth, called tusks, grow continually and become very sharp as a result of the lowers and uppers rubbing against each other.

Pigs are considered to be intelligent animals. Animal behavior expert Dr. Lyall Watson writes:

"Pigs have a proportionally larger brain than sheep, goats, or antelope and they also posses more of that part of the brain that governs the ability to reason. Pig brains are not unlike our own".

Pigs are social animals. In the wild, female pigs and their young live in extended family groups called "sounders." (Adult males are usually solitary.) The members of a sounder communicate with each other by sight, sound, and smell and cooperate to find food and to watch for, and sometimes, fight off, predators.

Pigs do not have effective sweat glands, so pigs cool themselves using water or mud during hot weather. They also use mud as a form of sunscreen to protect their skin from sunburn. Mud also provides protection against flies and parasites.

Pigs have a high reproductive rate. Females reach maturity in about a year and can have a litter of piglets, mostly 4 to 8 in the wild, every year after that. Pigs are unusual among hoofed animals in that the mother builds a nest in which to give birth and care for her young.

Pigs benefit the forest communities in which they live by clearing out dead animals; by keeping insects, which may be harmful to trees, under control; by breaking up the soil, which can promote plant growth; and by spreading the seeds of plants and the spores of fungi, including the truffle. However, on the other hand, feral pigs (domesticated pigs that are now wild) have also been described as agricultural pests and environmentally destructive. For example, the introduction of pigs to Australia resulted in destruction of habitat for native plants and animals, proliferation of weeds, destruction of pasture and crops, and environmental damage through rooting for food.

Pig farming is not only a profitable business but also a very popular and lucrative business. Pig is widely used to eat and pork. It takes a lot of time and money to make a profitable pig farm. To make a perfect pig farm everybody needs to follow some methods.

Pig Farming on Pasture

Firstly, it is very necessary to have a large area with a lot of grass and soil. Then it also needs a large fence so that the pigs can roam too far off easily and can feel comfort. But the farmers should always conscious about the structure of the fence. Farmers should use very strong wood and thick corner poles because the adult pigs become very strong and they can try to destroy the fence. As pigs are notorious diggers, farmers should dig deeply into the ground for poles. Thus farmers can farm pigs on the pasture.

Pig Farming in a Barn or Hut

Pig farming in a barn is generally easy. Farmers should use concrete to make a barn for pig farming. It is very necessary to keep the pigs inside the farm. Farmers should make the floor slope so that the water can be used to clean the barn easily. Farmers should separate the feeding area to the rest area for cleaning it easily. At least 10 feet deep and 10 feet wide area is perfect for every two pigs. The rest area should be half the size of the feeding area with a pool where they can bath and the pool should be at least 5 feet.

- Depending on number of pigs farmers should clean pens naturally twice in a week.
- It is very important to keep feeder full.
- Mud hole filled with water in outside pen.
- A lot of bedding is necessary in corner but not entire pen.

Selecting Pig Breeds

There are numerous pig breeds available throughout the world. You can choose the local breeds first which are easily available in your area. Some common and popular pig breeds are; Yorkshire, Spotted, Poland China, Landrace, Hampshire, Duroc, Chester White and Berkshire.

Feeding

Pig should be fed 18% crude protein. Farmers should not feed pig table scraps and garbage. For this farmers may be fall in danger. To farming, the pig should give the lettuce and other vegetables, but meat products are not appropriate. If the farmers give meat products the pig will put on too much fat and it will decrease the profit. Pig should feed 10-15 pounds of feed a day when it will 130 pounds heavier. To make a pig in a perfect weight, farmers should feed ½ cup of corn oil 2 times each day mixed with 1 egg 2 times each day and ½ cup of milk replacer powder twice a day,1 table spoon corn oil is also necessary. Proper and sufficient feeding brings success in pig farming.

Housing

To consider proper environment situation is very important to farming pig successfully. To farm pig farmers should make a pollution free environment. The pig industry needs a particular location. Without careful management of waste products, it may be very dangerous problem for child pig. It is a great and profitable process to make manure from the disposal of the pigs. It is very useful for agriculture. The environment of the outside and inside should protect all times because it is very essential to farming pigs. Inside environment is important for their health. The farming area should be made clean and dry. During the cold month, a heat lamp must be put and kept out of the north wind and south winds. Farmers should use straw as bedding during the winter. During summer season, it is very need that the pig has a place to lay in mud in the pen. If the pigs sweat or stay in a dry environment pigs will lose 5-10 pounds a day. So, always farmers should conscious about the pig's health and always try not to sweat them. To sell pigs and to get a lot of money pigs should reach at least 220-295 pounds of weight.

Health and Diseases

Farmers should conscious about pigs' health. So medical attention is essential.

- Farmers should be aware about the symptoms of diseases. Like no interest in food, diarrhea, eye discharge, excessive coughing, hernia, dry skin and irregular spots on skin, excessively long hair, back bone showing etc.
- If the pig has diarrhea, farmers should call veterinarian and treat it carefully. Without proper treat pigs can be dehydrated and lost its weights quickly. This situation can also lead to death.
- If the pigs are coughing it is very essential to contact with veterinarian as soon as possible.
- Pigs should not give antibiotics 21-45 days before slaughter date.
- Pigs must be wormed with safe guard swine wormier or Algard monthly to kill whip.

Sheep Farming

Sheep is a ruminant (cud-chewing) mammal of the genus Ovis. The sheep is usually stockier than its relative the goat; its horns, when present, are more divergent; it has scent glands in its face and hind feet; and the males lack the beards of goats. Sheep usually have short tails. In all wild species of sheep, the outer coat takes the form of hair, and beneath this lies a short undercoat of fine wool that has been developed into the fleece of domesticated sheep. Male sheep are called rams, the females ewes, and immature animals lambs. Mature sheep weigh from about 80 to as much as 400 pounds (35 to 180 kg).

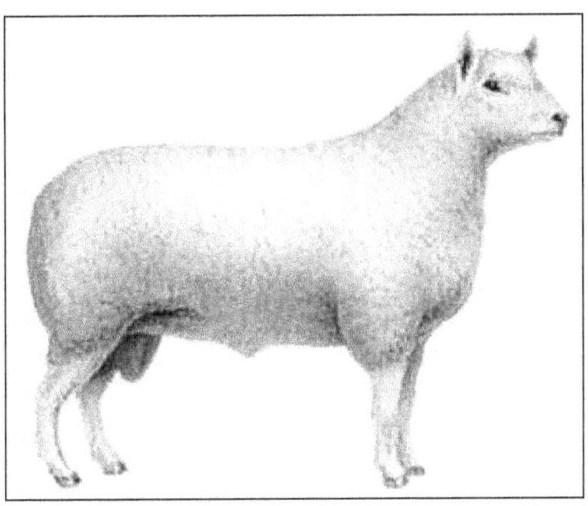

Cheviot ram.

Delaine ewe.

A sheep regurgitates its food and chews the cud, thus enabling its four separate stomach compartments to thoroughly digest the grasses and other herbage that it eats. The animals prefer grazing on grass or legume vegetation that is short and fine, though they will also consume high, coarse, or brushy plants as well. They graze plants closer to the root than do cattle, and so care must be taken that sheep do not overgraze a particular range. Sheep are basically timid animals who tend to graze in flocks and are almost totally lacking in protection from predators. They mature at about one year of age, and many breed when they reach the age of about one and a half years. Most births are single, although sheep do have twins on occasion. The lambs stop suckling and begin to graze at about four or five months of age.

Adult sheep with two lambs.

Sheep being herded.

Sheep were first domesticated from wild species of sheep at least 5000 BCE, and their remains have been found at numerous sites of early human habitation in the Middle East, Europe, and Central Asia. Domesticated sheep are raised for their fleece (wool), for milk, and for meat. The flesh of mature sheep is called mutton; that of immature animals is called lamb. There were estimated to be more than one billion sheep in the world in the early 21st century. The major national producers are Australia, New Zealand, China, India, the United States, South Africa, Argentina, and Turkey. Countries that have large areas of grassland are the major producers.

Domestic sheep differ from their wild progenitors and among themselves in conformation, quantity and quality of fleece, colour, size, milk production, and other characteristics. Most breeds of domesticated sheep produce wool, while a few produce only hair, and wild sheep grow a combination of wool and hair. Several hundred different breeds of sheep have been developed to meet environmental conditions influenced by latitudes and altitudes and to satisfy human needs for clothing and food. Breeds of sheep having fine wool are generally raised for wool production alone, while breeds with medium or long wool or with only hair are generally raised for meat production. Several crossbreeds have been developed that yield both wool and meat of high quality, however. Of the more than 200 breeds of sheep in the world, the majority are of limited interest except in local areas.

Merino ram.

Shropshire ewe.

Name	Type of Wool	Distribution	Characteristics	Comments
Black-Faced Highland, also called Scottish Blackface	Carpet	Originally Scotland, now also U.S., Italy, Argentina.	Black or mottled, horned.	Stylish appearance.
Columbia	Medium	Developed in U.S. since 1912.	Large, white-faced, hornless.	High wool yield; mutton acceptable.
Corriedale	Medium	Developed in N.Z., now also in U.S., Australia.	White-faced, hornless.	Bright, soft fleece; good quality lambs.
Cotswold	Long	Originally England, now also U.S.	Large, white-faced, hornless.	Coarse, curly fleece; acceptable mutton.
Dorset	Medium	Developed in England, now in U.K., U.S., Australia.	Medium-sized, white-faced.	Small wool yield; out-of-season lambs; horned and hornless varieties.
Hampshire	Medium	Developed in England, now also widespread in U.S.	Large, hornless; dark face and legs.	Superior mutton breed; limited wool.
Karakul	Fur	Originally Central Asia, now also Africa, Europe, U.S.	Medium-sized, fat-tailed.	Coats of very young lambs called Persian lamb.
Leicester	Long	Originally England, now U.K., North America.	Massive body, white-faced, broad-backed.	Heavy fleece.
Lincoln	Long	Originally England, now also Australia, N.Z., North and South America.	World's largest sheep, hornless.	Coarse, long wool is used chiefly for carpets.
Merino	Fine	Originally Spain, now also Australia, North America, South Africa.	Horned or hornless, heavily-wooled head.	Excellent, fine, soft fleeces.
North Country Cheviot	Medium	Originally Scotland, now widespread.	White chalk; large, deep-bodied.	Hardy; produces superior fleece.
Rambouillet	Fine	Developed in France from the 18th century, now also in U.S.	Smooth-bodied, horned or hornless.	Lambs mature rapidly; bred from Merino.
Romney	Long	Originally England, now also N.Z., North America, Australia.	Hornless with white face and legs.	Mostly raised for mutton; wool used for variety of products.

Southdown	Medium	Originally England, now also N.Z., Australia, North America.	Hornless with small, rounded body.	Raised for mutton; fleece is short.
Suffolk	Medium	Developed in England, now also in U.S.	Black face and legs, large, hornless.	Fine mutton breed; acceptable wool.

Sheep can be reared as free range (where there is no shortage of land) or under housing inside a shed. It is a very important component in dry land farming system. With very low investments can be made in to a profitable venture for small, marginal farmers and landless labours.

Sheep with its multi-facet utility for wool, meat, milk, skins and manure, form an important component of rural economy particularly in the arid, semi-arid and mountainous areas of the country. It provides a dependable source of income to the shepherds through sale of wool and animals.

The advantages of sheep farming are:

- Sheep do not need expensive buildings to house them and on the other hand require less labour than other kinds of livestock. The foundation stock is relatively cheap and the flock can be multiplied rapidly.
- Sheep are economical converter of grass into meat and wool.
- Sheep will eat varied kinds of plants compared to other kind of livestock. This makes them excellent weed destroyer.
- Unlike goats, sheep hardly damage any tree.
- The production of wool, meat and manure provides three different sources of income to the shepherd.
- The structure of their lips helps them to clean grains lost at harvest time and thus convert waste feed into profitable products.

Advantages of Sheep Rearing

- Well adopted to environment and poor management practices.
- The meat rate is increasing day by day.

- Sheep are suitable for wool and meat.
- Average of 1-2 kids per sheep per delivery.
- Average meat recovery of 22-30 kg/goat.
- Penning leads to manurial value to land.

Goat Farming

A goat is a member of the genus Capra of the bovid (Bovidae) family of even-toed ungulates, or hoofed mammals. There are several species of goats, all of them native to Asia, Europe, or northern Africa.

The domestic goat is descended from the wild goat, Capra aegagrus, and is sometimes considered a subspecies, C. aegagrus hircus, and sometimes a distinct species, C. hircus. It was one of the first animals domesticated by humans and remains an important domesticated animal today.

Goats provide numerous benefits to humans, including food (milk, meat, cheese), fiber and skin for clothing, brush and weed control, and as symbols in religion, folklore, and mythology. While the domestication of goats has been a tremendous benefit to humanity, poor management of goats has led to overgrazing of land and desertification in various regions. Properly managed goat herds can serve a valuable purpose in controlling weeds and in reducing excess undergrowth in forested areas vulnerable to fires.

A male goat is called a buck or billy, and a female is called a doe or nanny. Young goats are called kids.

The Rocky Mountain goat, Oreamnos americanus, of North America is not a true goat; although it, like sheep, the musk ox, the chamois, and other members of the goat-antelope subfamily (Caprinae), are closely related to the goats.

Biology and Habitat

Goats naturally live in rugged mountain or desert habitats. They are strong and skillful climbers and jumpers. Like other bovids, goats have a digestive system that enables them to survive on rough, low-quality plant material such as leaves and grasses. Goats can eat a wide variety of foods.

Goats are small for ungulates. Depending on the species, adults stand from 65 to 105 cm (2 to 3.5 feet) at the shoulder and weigh from 18 to 150 kg (40 to 330 lbs). Their bodies are covered with thick hair that protects them from the cold.

Both male and female goats have horns with the male's being larger. The horns are either curved or spiral shaped and can be as long as 165 cm (5.4 feet). Both male and female goats use their horns to fight among themselves and to fight off predators. Goats mostly live in groups ranging in size from 5 to 100 or so animals. Sometimes adult males live alone. The groups tend to keep moving, which helps them find food in their sparse habitats.

Goats are thought to be more intelligent than most other hoofed animals and seem to have a natural curiosity. They sometimes climb up into trees to feed on the leaves.

Goats give birth to one or two young after a gestation period of between 150 and 180 days, depending on the species. Like the young of most other bovids, new born goats can stand and follow their mothers almost as soon as they are born. The milk of goats is very rich and young goats grow rapidly. Mother goats are very protective of their young and will fight to defend them.

Species and Subspecies

Each of these goat species has several subspecies:

Markor

Kri-kri

- The wild goat, or Bezoar goat, Capra aegagrus, is found in Afghanistan, Armenia, Azerbaijan, Cyprus, Georgia, Greece, India, Iran, Iraq, Italy, Lebanon, Oman, Pakistan, Russia, Slovakia (where it was introduced), Syria, Turkey, and Turkmenistan.
- The domestic goat, C. aegagrus hircus or C. hircus, is found worldwide in the company of humans and in feral populations.
- The kri-kri or Cretan goat, C. aegagrus creticus, is found on the island of Crete and is thought to be descended from early domestic goats.
- The West Caucasian tur, C. caucasia, is found in Georgia and Russia.
- The East Caucasian tur, C. cylindricornis, is found in Azerbaijan, Georgia, and Russia.
- The markor, C. falconeri, is found in Afghanistan, India, Pakistan, Tajikistan, Turkmenistan, and Uzbekistan.
- The Alpine ibex, C. Ibex, is found in Austria, France, Germany, Italy, and Switzerland.
- The Nubian ibex, C. nubiana, is found in Egypt, Ethiopia, Israel, Jordan, Lebanon, Oman, Saudi Arabia, Sudan, Syria, and Yemen.
- The Spanish ibex, C. pyrenaica, is found in Portugal and Spain.
- The Siberian ibex, C. sibrica, is found in Afghanistan, China, India, Kazakhstan, Kyrgyzstan, Mongolia, Pakistan, and Russia.
- The Walia ibex, C. walie, is found in Ethiopia.

Nubian ibex

Domestication

Poor family with goat, painting.

Goats were one of the first animals domesticated by humans. This seems to have taken place first in the Middle East, perhaps as long as 10,000 years ago (the same time that sheep were also being domesticated). It has been suggested that the goats' natural curiosity and search for new food sources led them to associate with human settlements.

Keeping goats proved to be a valuable resource for early communities. They provided meat and milk, and their hair was used as fibre for clothing. The skin and the bones were also used. Historically, goat hide has been used for water and wine bottles, in both traveling and transporting wine for sale. It has also been used to produce parchment, which was the most common material used for writing in Europe until the invention of the printing press.

Baby domestic goat.

Domestic goats were generally kept in herds that wandered on hills or other grazing areas, often tended by goatherds who were frequently children or adolescents. These methods of herding are still utilized today. Goats can survive in difficult conditions. They also prefer different food than sheep and cattle, which are primarily grazers while goats are browsers, like deer, eating mostly leaves and leafy plants. Goats are better at fighting off predators than sheep and historically were kept sometimes with flocks of sheep to help defend the sheep.

Over time, goat keeping spread over most of Asia, Europe, and Africa. In parts of Africa and Asia, large herds of goats were maintained and land was often overgrazed. This has contributed to the expansion of deserts over large areas of these continents.

The Spanish and Portuguese brought goats to North and South America, and the English brought goats to Australia and New Zealand. Goats were also kept aboard ships to provide milk and meat on long voyages. Some of them were released by sailors on islands so that they could be hunted when the sailors returned. This has given rise to feral goat populations, which have caused much environmental damage on many islands around the world. Feral goats also exist on continents, but are not such an environmental problem there since their numbers are controlled by predators.

Human uses of Goats Today

Goats have continued to be an important domestic animal to the present day. The total number of domestic goats in the world is hard to estimate. China and India have the largest goat populations, over 100 million each, with most of them being raised for meat.

Brush Control

Domestic goats grazing.

Many farmers use inexpensive (i.e. not purebred) goats for brush control, leading to the use of the term "brush goats." (Brush goats are not a variety of goat, but rather a function they perform.) Because they prefer weeds (e.g. multiflora rose, thorns, small trees) to clover and grass, they are often used to keep fields clear for other animals. Their plant diet is extremely varied and includes some species that are toxic or detrimental to cattle and sheep. This makes them valuable for controlling noxious weeds and clearing brush and undergrowth. They will seldom eat soiled food or water unless facing starvation.

In efforts to reduce the environmental impact of human land use, some institutions, such as the NASA Ames Research Center in the heart of California's Silicon Valley, are turning to goats to reduce use of herbicides and mowing machines.

Meat

The taste of goat meat, called chevon, is said to be similar to veal or venison, depending on the age of the goat. It can be prepared in a variety of ways including stewed, baked, grilled, barbecued, minced, canned, or made into sausage. It is also healthier than mutton as it is lower in fat and cholesterol, comparable to chicken. It is popular in China, the Middle East, south Asia, Africa, Mexico, and north-eastern Brazil. Saudi Arabia is the largest importer of goat meat. It is not currently popular in Europe and the United States.

Some of the goat breeds bred for meat production are the Boer from South Africa, the Myotonic or Tennesse fainting goat from the United States, and the Kiko from New Zealand.

Milk and Cheese

Saanen does.

Goats' milk is more easily digested than cows' milk and is recommended for infants and people who have difficulty with cows' milk. The curd is much smaller and more digestible. Moreover it is naturally homogenized since it lacks the protein agglutinin. Furthermore, goats' milk contains less lactose, which means it will usually not trigger lactose intolerance in humans. Goats' milk is also used to make popular cheeses such as Rocamadour and feta.

Goat breeds bred as milk producers include the French-Alpine from France, the La Mancha from the United States, and the Oberhasli, Saanen, and Toggenburg from Switzerland.

Skin

Goat skin is still used today to make gloves, boots, and other products that require a soft hide. Kid gloves, popular in Victorian times, are still made today. The Black Bengal breed, native to Bangladesh, provides high-quality skin.

Fiber

Cashmere goats produce a fiber, "Cashmere wool," which is one of the best in the world. Cashmere fiber is very fine and soft, and grows beneath the guard hairs. Ideally there is a proportionally smaller amount of guard hair (which is undesirable and cannot be spun or dyed) to the cashmere fiber. Most goats produce cashmere fiber to some degree; however, the Cashmere goat has been specially bred to produce a much higher amount of it with fewer guard hairs.

Angora goat.

In south Asia, cashmere is called pashmina (Persian pashmina, meaning fine wool) and these goats are called pashmina goats (often mistaken as sheep). Since these goats actually belong to the upper Kashmir and Laddakh region, their wool came to be known as cashmere in the West. The pashmina shawls of Kashmir with their intricate embroidery are very famous.

The Angora breed produces long, curling, lustrous locks of mohair. The entire body of the goat is covered with mohair and there are no guard hairs. The locks can be six inches or more in length.

Goats do not have to be slaughtered to harvest the wool, which is instead sheared (cut from the body) in the case of Angora goats, or combed, in the case of Cashmere goats. The fiber is made into products such as sweaters. Both cashmere and mohair are warmer per ounce than sheep's wool and are not scratchy or itchy or as allergenic as wool sometimes is. Both fibers command a higher price than wool, compensating for the fact that there is less fiber per goat than there would be wool per sheep.

Pygmy goats in petting zoo.

Pets

Goats are sometimes kept as pets, although they require more care than dogs and cats. The Pygmy or West African dwarf from West Africa has become popular as a pet.

Goat farming is not a new enterprise. Generally goat farming means rearing goats for the purpose of harvesting milk, meat and fiber. At present, goat farming has become a profitable business and it requires a very low investment because of its multi-functional utility. Commercial goat farming business is contributing greatly to the economy and nutrition of a country. Goats are multi-functional animals. One can produce a wide variety of products from goats, such as milk, meat, fiber, manure etc. Goat's milk is used for producing full cream goat powder, skimmed goat milk powder, goat butter, goat milk cream, fresh goat milk etc. Goat meat is a great source of consumable meat which is very testy, nutritious and healthy. And goat's wool is being used in many purposes and skin of goat plays a vital role in leather industry.

Advantages of Goat Farming

There are many advantages of goat farming business. You can also raise goats along with your other livestock animals. Goats have been considered as poor man's cow (mini cow) for the poor people because of its immense contribution in rural economy and national income. Goat products like milk and meat is not only nutritious and easily digestible food but also a great source of regular income for the poor, landless and marginal farmers.

As goats are small sized animal, so they are easily maintained. Even they are easily maintained and cared by women and children. For successful goat farming business, you need to do some common tasks such as feeding, milking and caring. These simple tasks do not require much equipment, capital, labor or hard work. The main advantages of starting goat farming business are:

- Starting a goat farming business requires low initial investment or capital.
- Goats don't require huge area for housing because their body size is comparatively smaller than other livestock animals.
- Usually goats are very friendly in nature and very lovable.
- Goats are good breeders and they reach sexual maturity within their 7-12 months of age and give birth of kids within a short time. And some goat breeds produce numerous kids per kidding.
- Risks are less for goat farming (even in drought prone areas) than any other livestock farming business.
- Both male and female goats have almost equal value/price in the market.
- No religious taboo against goat farming and meat consumption.
- Goat meat and milk are cholesterol free and easily digestible.
- Goat milk is used for making various types of foods and it's very easy to digest than milk of cows.
- Commercial goat farming business has created a potential way of employment for unemployed people.
- Goats are multi-purpose animal. They can produce milk, meat, skin, fiber and manure at the same time.
- There is no need of a high end housing system for goats. Even they can easily share their living place with their owners or his/her other livestock animals.

- Goats are very suitable for mixed farming with other domestic animals.
- Diseases are less in goats than other domestic animals.
- Goats are easily available, comparatively cheaper in price, easy to maintain and always have a friendly disposition.
- They are capable of adopting themselves with almost all types of agro climatic environments or conditions. They can tolerate high and low temperature throughout the world and live happily. They also can tolerate hot climate more than other animals.
- According to the investment per unit they produce more than other domestic animals. And the ROI (return of investment) ratio is very good.
- Goats are generally smaller in size but reach slaughter age faster.
- Goats are called the “foster mother of human”. Because their milk is considered as the best milk for human consumption than any other species of livestock animal’s milk. And their milk is low cost, nutritious, wholesome and easily digestible. All aged people from child to old one can easily digest goat’s milk. Goat milk also has lesser allergic problems. And used as an Ayurvedic medicine for the people who are ailing with diabetes, asthma, cough etc.
- Goat milk is suitable for preparing various types of milk products. And goat meat has a huge demand and high price in the local and international markets. Even you can consider exporting your products to the foreign countries for more profits.
- Goats can be milked as often as required. This also prevent refrigeration costs and milk storage problems.
- You can use the goat’s manure as a high quality natural fertilizer in crop field. This will directly help to maximize crop production.
- As goat farming business is very profitable, so many govt. and non govt. banks are providing loans for starting this business.
- This business requires less labor and you can easily use your family labor for raising goats.
- Commercial goat farming business is a great source of employment and income. So unemployed educated people can easily create a great employment and income source through raising goats commercially.
- Along with the above advantages, there are also many advantages of raising goats commercially.

Feeding Goats

Well goat feed management is also very essential for commercial goat farming business. Goats are not carnivorous animal. They don’t eat other animals. Usually goats prefer to eat grasses, plants, shrubs, weeds and herbs. Besides those feeds, goat also need energy, portion, vitamins, fiber and water for proper growth and for making better profits from this business. It will be better if you have sufficient knowledge about what to feed goats.

Care and Management

Always try to take good care of your goats. Never feed them contaminated food or polluted water. Keep their house as much neat and clean as possible. And clean their house on a regular basis. For commercial production, keep kids, bucks and does separated for each other. Take extra care to the breeding bucks, kids and pregnant does. Keep the kids with their mother for several weeks after their birth.

Avoid using same buck for mating with numerous does at same day. Artificial insemination is also a great way for breeding your does. Vaccinate them timely, for keeping them free from all types of diseases and health problems. If possible, stock some necessary vaccines and medicines and keep good relation with a veterinary doctor.

Vaccination

Various types of viral diseases like PPR, goat pox, foot and mouth diseases and bacterial diseases like anthrax, brucelosis etc. are very harmful for goats. So, proper vaccination is a must to prevent these types of diseases. The does which was not vaccinated PPR, goat pox, brucellosis vaccines previously, vaccinate them at the fifth month of gestation period. Vaccinate the kids PPR vaccine when they reach 5 months of age. Always take good care of your animals and vaccinate them timely to prevent unwanted health hazard and diseases.

Goat Vaccination		
Vaccine Name	Applying Rate	Applying Method
PPR	1 ml	Injection Under Skin
Foot and Mouth Disease	2 ml	Injection Under Skin
Anthrax	1 ml	Injection Under Skin

Horse Farming

Horse, (Equus caballus) is a hoofed herbivorous mammal of the family Equidae. It comprises a single species, Equus caballus, whose numerous varieties are called breeds. Before the advent of mechanized vehicles, the horse was widely used as a draft animal, and riding on horseback was one of the chief means of transportation. Horses were among the last species of livestock to be domesticated.

Form and Function

A mature male horse is called a stallion, the female a mare. A stallion used for breeding is known as a stud. A castrated stallion is commonly called a gelding. Formerly, stallions were employed as riding horses, while mares were kept for breeding purposes only. Geldings were used for work and as ladies' riding horses. Recently, however, geldings generally have replaced stallions as riding horses. Young horses are known as foals; male foals are called colts and females fillies.

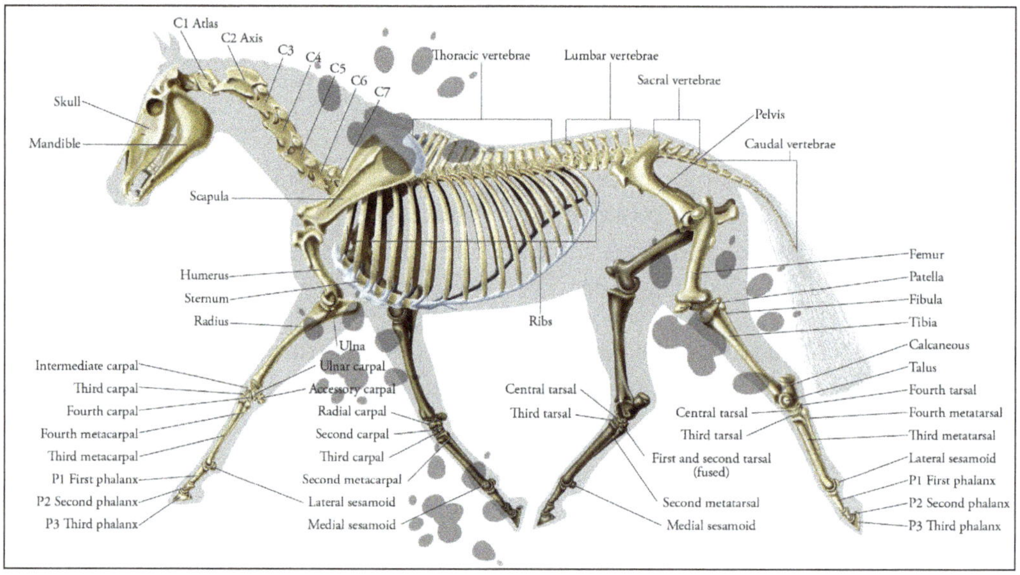

Horse anatomy. Skeleton (top) and external structure (bottom) of the horse.

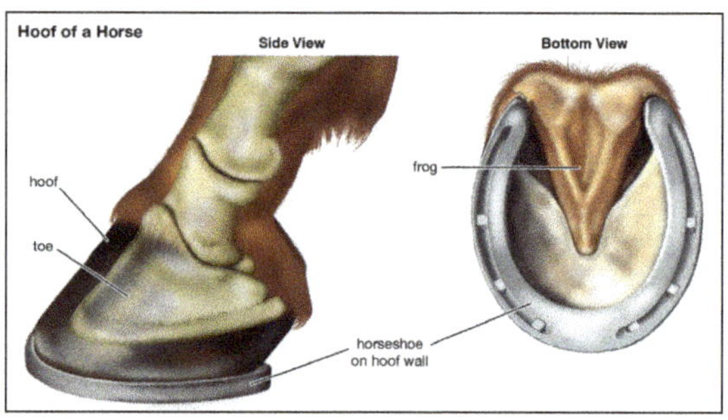

Hoof of a horse.

Anatomical Adaptations

The primitive horse probably stood 12 hands (about 120 cm, or 48 inches [1 hand = 10 cm, or 4 inches]) tall at the withers, the high point on the back at the base of the neck, and was dun coloured (typically brownish to dark gray). Domestic horses gone wild, such as the mustangs of western North America, tend to revert to those primitive features under random mating: they generally are somewhat taller (about 15 hands [152.4 cm, or 60 inches]), are usually gray, dun, or brownish in colour, and move in herds led by a stallion.

The horse's general form is characteristic of an animal of speed: The long leg bones pivot on pulley-like joints that restrict movement to the fore and aft, the limbs are levered to muscle masses in such a way as to provide the most efficient use of energy, and the compact body is supported permanently on the tips of the toes, allowing fuller extension of the limbs in running.

The rounded skull houses a large and complex brain, well developed in those areas that direct muscle coordination. While the horse is intelligent among subhuman animals, it is safe to say that the horse is more concerned with the functioning of its acute sensory reception and its musculature than with mental processes. Though much has been written about "educated" horses that appear to exhibit an ability to spell and count, it is generally agreed that in such cases a very perceptive animal is responding to cues from its master. But this ability is remarkable enough in its own right, for the cues are often given unconsciously by the human trainer, and detection of such subtle signals requires extremely sharp perception.

The horse, like other grazing herbivores, has typical adaptations for plant eating: a set of strong, high-crowned teeth, suited to grinding grasses and other harsh vegetation, and a relatively long digestive tract, most of which is intestine concerned with digesting cellulose matter from vegetation. Young horses have milk (or baby) teeth, which they begin to shed at about age two and a half. The permanent teeth, numbering 36 to 40, are completely developed by age four to five years. In the stallion these teeth are arranged as follows on the upper and lower jaws: 12 incisors that cut and pull at grasses; 4 canines, remnants without function in the modern horse and usually not found in mares; 12 premolars and 12 molars, high prisms that continue to grow out of the jaw in order to replace the surfaces worn off in grinding food.

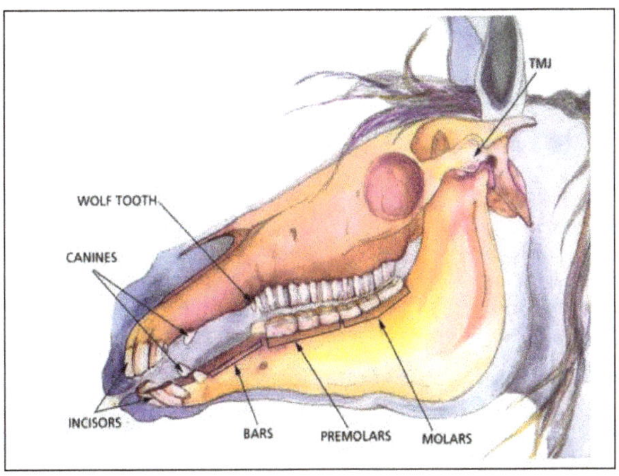

Teeth of a horse.

Under domestication the horse has diversified into three major types, based on size and build: draft horses, heavy-limbed and up to 20 hands (200 cm, or 80 inches) high; ponies, by convention horses under 14.2 hands (about 147 cm, or 58 inches) high; and light horses—the saddle or riding horses—which fall in the intermediate size range. Domestic horses tend to be near-sighted, less hardy than their ancestors, and often high-strung, especially Thoroughbreds, where intensive breeding has been focused upon speed to the exclusion of other qualities. The stomach is relatively small, and, since much vegetation must be ingested to maintain vital processes, foraging is almost constant under natural conditions. Domestic animals are fed several (at least three) times a day in quantities governed by the exertion of the horse.

Senses

The extremely large eyes placed far back on the elongated head admirably suit the horse for its chief mode of defense: flight. Its long neck and high-set eyes, which register a much wider range than do the eyes of a human being, enable the horse to discern a possible threat even while eating low grasses. Like human vision, the horse's vision is binocular, but only in the narrow area directly forward. Evidence suggests that a horse's vision is limited in its ability to register colour; horses can detect yellow and blue but not red and green. While visual acuity is high, the eyes do not have variable focus, and objects at different distances register only on different areas of the retina, which requires tilting movements of the head. The senses of smell and hearing seem to be keener than in human beings. As the biologist George Gaylord Simpson put it in Horses:

"Legs for running and eyes for warning have enabled horses to survive through the ages, although subject to constant attack by flesh eaters that liked nothing better than horse for supper".

Colour and Pattern

From the dun of the primitive horse has sprung a variety of colours and patterns, some highly variable and difficult to distinguish. Among the most important colours are black, bay, chestnut (and sorrel), palomino, cream, and white.

In figure, common horse colours: Dappled gray (top left), dun (centre left), brown (bottom left), strawberry roan (top centre), chestnut (centre), skewbald (a type of pinto, bottom centre), palomino (top right), bay (centre right), black (bottom right).

The black colour is a true black, although a white face marking (blaze) and white ankles (stockings) may occur. The brown horse is almost black but has lighter areas around the muzzle, eyes, and legs. Bay refers to several shades of brown, from red-brown and tan to sandy. Bay horses have a black mane, tail, and (usually) stockings. There is a dilution (or lightening) gene—called silver or silver dapple—that mainly influences the dark colours of the coat. Chestnut is similar to bay but with none of the bay's black overtones. Lighter shades of chestnut are called sorrel. The palomino horse runs from cream to bronze, with a flaxen or silvery mane and tail. The cream is a diluted sorrel, or very pale yellow, nearly white. White in horses is variable, ranging from aging grays to albinos with blue eyes and pink skin and to pseudoalbinos with a buff mane or with brown eyes. The chief patterns of the white horse are gray, roan, pinto, sabino, and appaloosa. Gray horses are born dark brown or black and develop white hairs as they age, becoming almost all white in advanced years. Roan refers to white mixed with other colours at birth: blue roan is white mixed with black; red roan is mixed white and bay; and strawberry roan is white and chestnut.

The pinto is almost any spotted pattern of white and another colour; other names, such as paint, calico, piebald, skewbald, overo, and tobiano, refer to subtle distinctions in type of colour or pattern. Appaloosa (leopard complex) is another extremely variable pattern, but the term generally refers to a large white patch over the hips and loin, with scattered irregular dark spots.

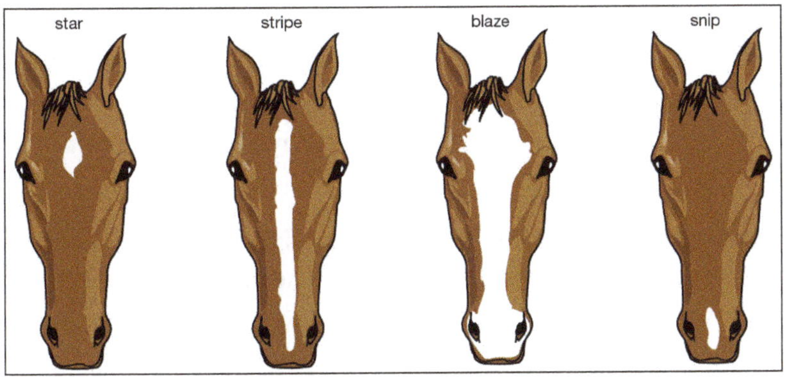

Different types of horse facial markings.

Studies of five coat-colour genes in DNA samples from ancient, pre-domesticated horses have shown that these horses predominantly carried the genes for black or bay. Scientists believe that it is very likely that these horses also carried the dun dilution gene. The leopard (Appaloosa) mutation was also discovered, which was found to be consistent with some cave paintings dating to 25,000 years ago that depict spotted horses. Mutations for chestnut, tobiano, and sabino were also observed and were dated to 3,000 years ago, whereas the buckskin variant had emerged by about 1,000 years ago. Most of the variation in coat colour appeared after domestication occurred and was likely the result of artificial selection by humans.

Pinto with tobiano pattern.

Behaviour

The horse's nervous system is highly developed and gives proof to varying degrees of the essential faculties that are the basis of intelligence: instinct, memory, and judgment. Foals, which stand on their feet a short while after birth and are able to follow their mothers within a few hours, even at this early stage in life exhibit the traits generally ascribed to horses. They have a tendency to flee danger. They express fear sometimes by showing panic and sometimes by immobility. Horses rarely attack and do so either when flight is impossible or when driven to assault a person who has treated them brutally.

Habit governs a large number of their reactions. Instinct, together with a fine sense of smell and hearing, enables them to sense water, fire, even distant danger. An extremely well-developed sense of direction permits the horse to find its way back to its stables even at night or after a prolonged absence. The visual memory of the horse prompts it to shy repeatedly from an object or place where it had earlier experienced fear. The animal's auditory memory, which enabled ancient army horses or hunters to follow the sounds of the bugles, is used in training. When teaching, the instructor always uses the same words and the same tone of voice for a given desired reaction. Intelligent horses soon attach certain movements desired by their trainers to particular sounds and even try to anticipate their rider's wishes.

While instinct is an unconscious reaction more or less present in all individuals of the same species, the degree of its expression varies according to the individual and its development. Most horses can sense a rider's uncertainty, nervousness, or fear and are thereby encouraged to disregard or even deliberately disobey the rider.

Cunning animals have been known to employ their intelligence and physical skill to a determined end, such as opening the latch of a stall or the lid of a chest of oats.

Reproduction and Development

The onset of adult sex characteristics generally begins at the age of 16 to 18 months. The horse is considered mature, depending on the breed, at approximately three years and adult at five. Fecundity varies according to the breed and may last beyond age 20 with Thoroughbreds and to 12 or 15 with other horses. The gestation period is 11 months; 280 days is the minimum in which the foal can be born with expectation to live. As a rule, a mare produces one foal per mating, twins occasionally, and triplets rarely. The foal is weaned at six months.

The useful life of a horse varies according to the amount of work it is required to do and the maintenance furnished by its owner. A horse that is trained carefully and slowly and is given the necessary time for development may be expected to serve to an older age than a horse that is rushed in its training. Racehorses that enter into races at the age of two rarely remain on the turf beyond eight. Well-kept riding horses, on the contrary, may be used more than 20 years.

The life span of a horse is calculated at six to seven times the time necessary for its physical and mental development—that is, 30 to 35 years at the utmost, the rule being about 20 to 25 years. Ponies generally live longer than larger horses. There are a number of examples of horses that have passed the usual limit of age. The veterinary university of Vienna conserves the skeleton of a Thoroughbred mare of 44 years of age. There have been reports made of horses living to their early 60s in age.

Nutrition

The horse's natural food is grass. For stabled horses, the diet generally consists of hay and grain. The animal should not be fed immediately before or after work, to avoid digestive problems. Fresh water is important, especially when the horse is shedding its winter coat, but the animal should never be watered when it is overheated after working. Oats provide the greatest nutritional value and are given especially to foals. Older horses, whose teeth are worn down, or those with digestive troubles, can be provided with crushed oats. Chaff (minced straw) can be added to the oat ration of animals that eat greedily or do not chew the grain properly. Crushed barley is sometimes substituted in part for oats. Hay provides the bulk of the horse's ration and may be of varying composition according to locale. Mash is bran mixed with water and with various invigorating additions or medications. It may be given to horses with digestive troubles or deficient eating habits.

Corn (maize) is used as a fattening cereal, but it makes the horse sweat easily. Salt is needed by the horse at all times and especially when shedding. Bread, carrots, and sugar are tidbits often used by the rider or trainer to reward an animal. In times of poverty, horses have adapted to all sorts of food—potatoes, beans, green leaves, and in Iceland even fish—but such foods are not generally taken if other fare is available. A number of commercial feed mixes are available to modern breeders and owners; these mixes contain minerals, vitamins, and other nutrients and are designed to provide a balanced diet when supplemented with hay.

Feeding

The specific and exact nutrient requirements of horses are poorly understood. Usually, these may be supplied economically from pasture forage, harvested roughages, and concentrates. Good quality grass-legume pastures, in addition to iodized or trace-mineralized salt, will supply adequate nutrients to maintain an adult horse at light work (such as pulling a small cart) or mares during pregnancy. Lush, early spring pasture is very high in water and protein contents and may need to be supplemented with a high-energy source, such as grain, to meet the needs of horses performing medium to heavy work (such as plowing).

Conversely, late fall- and winter-pasture forage is low in water and protein and may require protein and vitamin A supplementation. High-quality legume hays, such as early bloom alfalfa, are preferred for horses, especially those that are growing or lactating. Moldy or dusty feeds should be avoided because horses are extremely susceptible to forage poisoning and respiratory complications. Grass hays, such as timothy, prairie grass, orchard grass, and bluegrass, were preferred by early horsemen, especially for race horses, because they were usually free from mold and dust and tended to slow down the rate of passage through the intestinal tract. These hays are low in digestible energy and protein, however, and must be adequately supplemented. Silages of all sorts should be avoided since horses and mules are extremely susceptible to botulism and digestive upsets.

Oats are the preferred grain for horses because of their bulk. Corn (maize), barley, wheat, and milo can be used, however, whenever they are less expensive. Weanling foals require three pounds of feed per hundred pounds of live weight per day; as they approach maturity, this requirement drops to one pound of feed per hundred pounds of live weight daily. Horses normally reach mature weight at less than four years of age and 80 percent of their mature weight at less than two years of age.

A large and ever-growing number of horses stabled in cities and suburbs where sufficient roughages cannot be grown provide a large market for complete horse rations, including roughage, which are tailored to the total needs of specific animals according to their particular function at a given time, such as growth, pregnancy, lactation, or maintenance.

Horses will vary from the normal requirement in terms of weight, temperament, and previous nutrition. Foals will eat some pasture grass, forage, or hay when they are three days old and grain when they are three weeks old.

Management

Highly bred light horses are notoriously poor reproducers. Many horse farms consider a 60 percent foaling rate for a large band of mares to be average. Most large horse farms employ resident veterinarians to check for abnormalities or disease before breeding and to check mares for pregnancy 40 to 45 days after breeding. Because many mares conceive only every other year, expert assistance at foaling time is an absolute necessity, especially if the foal is sired by an expensive stallion out of a valuable mare.

The gestation period of horses is 340 days. The period between estrus ranges from 18 to 28 days with an average of 22 days. Mature stallions can safely mate with from 50 to 100 mares per season, although the practice with expensive Thoroughbreds is to book no more than 35 to 40 mares.

The feet and legs of horses demand unusual attention. The old adage "no foot no horse" remains apt. Hooves should be trimmed regularly, beginning when the horse is a foal or only a few months old. Otherwise they may grow long and uneven, causing improper action, undue stress on joints, and broken or cracked hooves. Horses that are worked regularly, especially on hard and stony ground, as well as show horses and race horses in service, must be shod. Shod horses should have their hooves trimmed and their shoes refitted every four to six weeks. Tendency toward unsoundness is probably inherited but may be aggravated by poor hoof care and excessive stresses.

Diseases

Horses are especially susceptible to tetanus or lockjaw but can be given two-year protection through the use of a commonly accepted toxoid. There are two common types of abortion in horses: virus abortion, specifically viral rhinopneumonitis, and the Salmonella type. The former, which produces an influenza with pinkeye, catarrh, general illness, and abortion, affects both mares and foals, but all surviving horses develop natural resistance soon after infection. Pregnant mares thought to be subjected to infection may be given some protection by available vaccines. The Salmonella type of abortion can be prevented completely by vaccination. Encephalomyelitis, or sleeping sickness, is prevented by vaccination.

A specific vaccine is available for anthrax, which is prevalent in Asia. Hemolytic anemia of foals has become a problem. Foals so afflicted are born normal but soon become sluggish and progressively weaker; the membranes of their eyes, mouth, and lips become very pale and the heartbeat becomes rapid. This condition is caused by antibodies in the mare's milk that destroy the foal's red blood cells. These antibodies are caused by the difference in blood type between the foal and the mother. Newborn foals can be muzzled to avoid nursing while their blood is checked for reaction against the serum and milk of its mother. Where reactions are noted, the mare is hand-milked at hourly intervals for 12 to 24 hours, and the foal is fed milk from another suitable mare or a milk substitute. Horses are quite susceptible to various infections, but rotation of pastures, strict sanitation, and the use of suitable vermifuges are quite effective.

Poultry Farming

Poultry, in animal husbandry refers to birds raised commercially or domestically for meat, eggs, and feathers. Chickens, ducks, turkeys, and geese are of primary commercial importance, while guinea fowl and squabs are chiefly of local interest.

Poultry farming is defined as a term for rearing and keeping of birds such as fowl, duck and hen for egg and meat. Poultry farming has become popular because it is comparatively easy to start and maintain. It gives quick return within one to six month of investments, is easily manageable and requires less space and labour. Poultry birds and their eggs are a rich source of nutrients.

Poultry Housing

Poultry production systems should provide fresh air, clean feed and water, protection against predators, shelter from cold, rain, wind, sun and excessive heat; as well as a source of heat when birds are young. Basically, the birds need to be able to grow, sleep and lay eggs in comfort. The birds should also be free from stress and disease. The basic requirements for poultry housing are:

1. Protection from weather.
2. Protection from predators.
3. Enough space.
4. Adequate ventilation.
5. A clean environment.
6. Access to dust bathing facilities.

Space requirements, or density of birds per unit area: This is the most important basic principle in housing, as the space available determines the number and type of poultry that can be kept.

Minimum Space Requirements for Different Poultry

Type of bird	Sq.ft/ Bird Inside	Sq.ft/ Bird Outside Runs
Bantam Chickens	1	4
Laying Hens	1.5	8
Quail	1	4
Phesant	5	25
Ducks	3	15

Geese	6	18
Minimum Requirement of chicken for floor and perch space		
Chicken Types	Floor Space (ft2/ bird)	Perch space (per bird)
Layer	3.6	25cm(10")
Dual Purpose	2.7	20cm(8")
Meat	2.1-2.7	15-20cm(6-8")

System of Poultry Housing

Deep Litter System

Birds are fully confined within a house (3 to 4 birds/ m2) but can move around freely. The floor is covered with a deep litter (5 to 10 cm deep layer) of grain husks (maize or rice), straw, wood shavings or a similarly absorbent, non-toxic material.

The fully enclosed system protects the birds from thieves and predators and is suitable for specially selected commercial breeds of egg or meat producing poultry (layers, breeder flocks and broilers).

Slatted Floor Systems

Wire or wooden slatted floors are used instead of deep litter, which allow stocking rates to be increased to five birds/m^ of floor space. Birds have reduced contact with faeces and are allowed some freedom of movement. Faeces can be collected from below the slatted floor and used as fertilizer.

Battery Cage Systems

This is usually used for laying birds, which are kept throughout their productive life in small cages. There is a high initial capita investment, and the system is mostly confined to large-scale commercial egg layer operations.

Broiler Management of Poultry

1. Since broilers are being reared for meat it is important that they always have an adequate supply of high quality broiler feed.
2. As the birds get heavier, they will need more floor space and ventilation.
3. It may be well to use night lights equivalent to 15 watts per 200 square feet. This allows birds to eat at night and helps prevent pile-ups.
4. Keep litter dry to help prevent breast blisters. Provide ample cool, clean water.

Pullet Management of Poultry

1. Feed starter and grower feeds.
2. Keep young and old birds separated. If it is necessary for one person to care for young and old birds, care for the young birds first each day.

3. Remove any unthrifty pullets.
4. Keep birds free of parasites.
5. Keep complete and accurate records.

Layer Management of Poultry

1. Clean and disinfect laying house before placing pullets in it.
2. If floor management is used, put in 4 to 6 inches of clean litter.
3. House only well developed well fleshed pullets.
4. Use artificial light to provide 14 hours of total light per day – one 40 watt bulb per 200 square feet, hung 8 feet above the floor.
5. Use a feeding program.
6. Keep birds free of parasites.
7. Keep complete and accurate records.
8. Remove obvious culls.

Brooding of Poultry

The care and management of young chicks during early part of their life is called brooding i.e. up to the age of 3-5 weeks.

Brooders

1. Hover-type – Follow manufacturer's direction. Starting temperature at the edge of the hover should be 95 °F. for the first week and reduced 5 °F each week.
2. Heat lamp – Use either white or infra-red heat lamps.
3. Home made brooder – An inexpensive method of supplying heat to a few chicks is to place a 100-watt light bulb inside a gallon tin can and place the can on the floor of the brooder house.

Cages for Poultry

Normally cages are 18 inches deep and 16 inches high in the following widths:

1. 8 inches wide – one bird per cage.
2. 10 inches wide – can accommodate two layers per cage.
3. 12 inches wide – can accommodate two or three layers per cage.
4. Although more than one bird per cage can be housed, one bird per cage is recommended for small flocks for the following reasons; cleaner eggs, fewer cracked eggs, usually higher

production, better bird plumage and less cannibalism. Manure management is much easier since the manure is not so concentrated, thus aiding in drying and reducing odors and fly problems.

Hatching Egg Production and Care

- Keep 1 male for each 10-12 females.
- Males should be kept with females at least 1 week prior to saving hatching eggs to insure high egg fertility.
- Feed a complete diet.
- Collect eggs at least 3 limes per day.
- The hatching of fertilized eggs requires about 21 days and hatching takes place at 38°C temperature.

Incubation Essentials

- Obtaining a small incubator: Still air incubator can be purchased or constructed. Small forced-air self-turning incubators are commercially available.
- Proper operating temperature: A still air incubator requires an operating temperature of 102 to 103°F. At a position level with the top of the eggs. A forced draft (which contains a fan for circulating the air) incubator should be operated at 99 to 100°F. Do not place the incubator in drafts or direct sunlight which may cause extreme fluctuations in temperature.
- Sufficient humidity: Wet bulb reading of 86°F. For a small incubator, moisture can be added to the air by placing a small pan of water under the egg tray. It may be necessary to sprinkle the eggs lightly with warm water at the time of hatching to prevent the chicks from sticking to the shell.
- Turning of eggs: Eggs should be turned an odd number of times and a minimum of three times each day. Mark each egg as an aid in determining that all eggs are turned from one side to the other at each turning. For self- turning incubation, follow manufacturer's instructions.

An Egg contains 67% water, 13% protein, 9% fat and 11% minerals.

Moulting

It is a natural physiological process for the birds to renew old feathers at the end of the first year of lying.

Dubbing

Removal of comb may be restricted to in day old chicks belonging to breeds, which have larger/ lopped comb.

Debeaking

Debeaking is cutting off part of the upper beak. It helps in preventing peaking injuries and cannibalism among chicks.

Poultry Disease

Rearing of poultry birds requires properly ventilated place and vaccination of new born chicks. Poultry diseases can be classified as infectious or non-infectious. Non-infectious diseases are caused by faulty management, faulty feed preparation and inadequate diet or nutritionally efficient disease.

Disease	Casual Organism
Bacterial Disease:	
Salmonellosis	Salmonella
Fowl Coryza	Hemophilus paragallinarum
Fowl Cholera	Pasteurella Multocida
Pullorum Disease	Salmonella pullorum
Viral Disease:	
Infectious Bronchitis	IB Virus
Turkey Rhinotracheitis/Swollenhead syndrome	Pneumovirus
Marek's Disease/ Fowl paralysis	Herpes virus or MD virus
Fowl Pox/Avian Pox/Avian Diphtheria	Fowl Pox Virus
Ranikhet (RD)/New Castle disease	Paramyxoviris-1
Protozoan disease:	
Coccidiosis	Eimeria acervulina
Fungal disease	Aspergillosis, Moniliasis
ParasiticDisease	Likce, Mites, Tick and fleas(external) and round worms, tapeworm and hexamitiasis (internal)

Vaccination Schedule for poultry		
Age of Bird	Name of Vaccine/disease	Dose/route
1st day	HVT/Marek's Disease vaccine	0.2ml s.c. in neck.
7th Day	RD Vaccine/B1/NDF/ Clone 30 (Ranikhet disease)	One drop in eye or nose.
12th Day	IB disease vaccine 'strain 228'	One drop in eye .
21st Day	HVT/ Marek's Disease vaccine To be repeated	0.2ml s.c. in neck.
21-25th Day	IB vaccine	One drop in eye .
8th Weeks	Fowl Pox Vaccine	0.2ml i.m. or one drop by pin prick method.
9-11th Weeks	R2B Vaccine/ Ranikhet disease	0.5ml i.m. or s.c. in wing muscle.
12th Weeks	IB Vaccineto be repeated	In drinking water.
14th weeks	Fowl pox vaccine To be repeated	0.2ml i.m. or one drop by pin prick method.

References

- Livestock: newworldencyclopedia.org, Retrieved 2 March, 2019
- Livestock-farming: britannica.com, Retrieved 11 May, 2019

- Cattle-livestock, animal: britannica.com, Retrieved 20 August, 2019
- Cattle-farming-animal-husbandry, biology: byjus.com, Retrieved 24 January, 2019
- Pig-farming: roysfarm.com, Retrieved 4 March, 2019
- Sheep, animal: britannica.com, Retrieved 7 July, 2019
- Animhus-sheep-Rearing, animal-husbandry: tnau.ac.in, Retrieved 17 February, 2019
- Goat-farming: roysfarm.com, Retrieved 8 April, 2019
- Horse, livestock-farming: britannica.com, Retrieved 10 January, 2019
- Poultry-farming-meaning-poultry-housing-and-brooding, farming: biologydiscussion.com, Retrieved 13 June, 2019

Chapter 3

Animal Reproduction and Breeding

The biological process through which new animals are produced by their parents is termed as animal reproduction. Animal breeding refers to the controlled propagation of domesticated animals for the purpose of improving desirable qualities. This chapter discusses in detail animal reproduction as well as the different techniques used for animal breeding.

Animal Reproduction

In order for a species to thrive, it must be able to reproduce itself. Producing offspring requires the proper functioning of both the male and female reproductive systems, each of which consists of several parts that have specific purposes in the reproductive system. Hormones also play an important role in reproduction. For example, in the male reproductive system, the parts and hormones work together to produce sperm and transfer it to the female.

Parts of the Reproductive System

The male animal possesses special reproductive organs. These organs are similar in all male mammals, although some differences in form and function do exist. The bull can be used to illustrate the male mammalian reproductive system.

Bull – A bull is a male bovine. Its reproductive system contains a number of different parts. A bull has two testicles, which are located in the scrotum beneath the animal. An epididymis is located in the scrotum next to each testicle. The epididymis is connected to the urethra by the vas deferens. The seminal vesicles, prostate gland, and Cowper's gland (also called the bulbourethral gland) are located next to the bladder where the vas deferens meets the urethra. The urethra extends from this point to the penis. The sigmoid flexure is located midway along the urethra, which also has a retractor muscle attached to it. The sheath provides an opening for the penis.

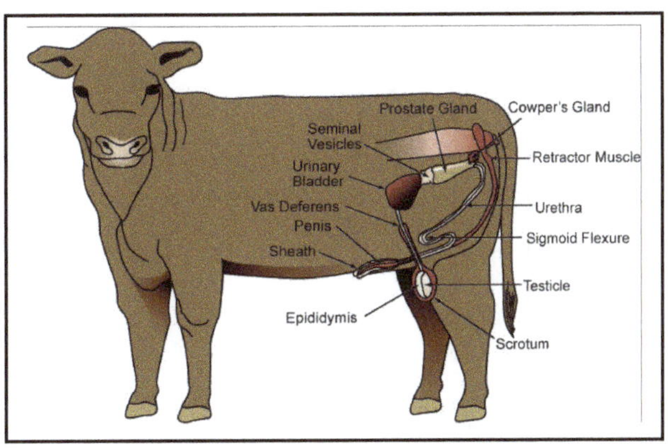

Reproductive System of the Bull.

Boar – A boar is a male swine. The boar's reproductive parts are pictured. One difference between the boar and the bull is the positioning of the testicles and scrotum, which are located at the rear of the boar rather than at the bottom of the animal. The form of the penis, which is shaped like a corkscrew, also differs from the bull's.

Ram – A ram is a male sheep. The ram differs from the bull in that a filiform appendage extends from the head of the penis. In the ram, the sheath is also referred to as a prepuce.

Stallion – In horses, a male is called a stallion. In contrast to the bull, the stallion's scrotum is located further toward the rear of the animal and is less pendulous. The stallion also lacks a sigmoid flexure.

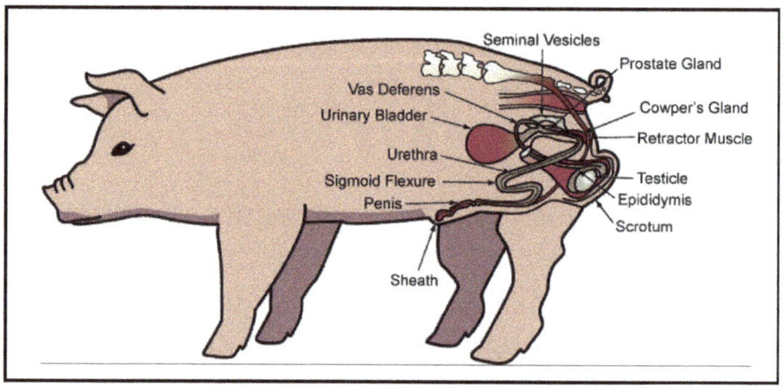

Fowl – Fowl include chickens, turkeys, and other birds. Unlike other animal species covered in this lesson, fowl are not mammals. Thus, they have a very different reproductive system from the others, as illustrated in Figure below. Fowl do not have a scrotum. Instead, their testicles are located within the abdomen, next to the backbone. The vas deferens connects the testicles to the cloaca, papillae, and vent. The papillae are small, finger-like projections within the cloaca. Fowl do not have a urethra or urinary bladder.

Functions of the Parts

Each part has a specific function in the reproductive system. If any of the parts do not function correctly, the animal may have difficulty with reproduction or be unable to reproduce. Following is a description of the functions of each of the reproductive parts of the bull.

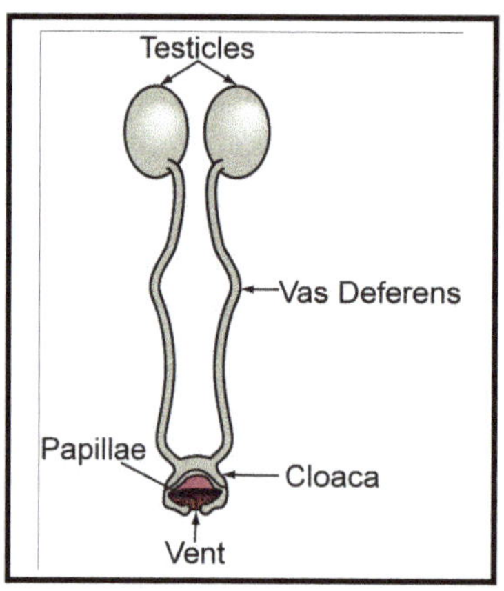

Reproductive System of a Rooster.

Testicles – The testicles are the primary reproductive part. They produce hormones associated with reproduction as well as the male gametes, or sex cells, called sperm. The testicles are made up of seminiferous tubules, which are tiny coiled tubes that produce the sperm, as well as interstitial cells located between the tubules that secrete reproductive hormones.

- Scrotum – The scrotum is a sac or pouch of skin that carries the testicles and regulates their temperature. When cold, the testicles are drawn close to the body for warmth. In hot weather, the testicles hang away from the body.
- Epididymis – The epididymis is a tube with three parts: a head, body, and tail. Sperm is stored in the epididymis for maturation. It also concentrates the sperm and transports it from the testicles to the vas deferens.
- Vas deferens – The vas deferens is a tube that transports sperm from the epididymis to the urethra.
- Urethra – The urethra carries sperm and urine to the penis.
- Urinary bladder – The bladder stores urine before it is released to the urethra. It has no reproductive function.
- Seminal vesicles – The seminal vesicles produce the seminal fluid that transports and protects the sperm.
- Prostate gland – The prostate gland releases fluid that mixes with the seminal fluid and nourishes the sperm. The combination of the sperm with the fluids from the seminal vesicles and prostate gland is referred to as semen.
- Cowper's (bulbourethral) gland – This gland releases a fluid into the urethra that cleanses and neutralizes it to allow the sperm to survive the passage to the penis. The secretions pass through the urethra prior to the semen.
- Sigmoid flexure – This S-curved muscle extends the penis outside the body during mating.
- Retractor muscle – This muscle pulls the penis back into the body.
- Penis – The penis is an organ that deposits semen into the female reproductive tract. It also excretes urine from the body.
- Sheath – The sheath is a fold of skin that covers and protects the penis when it is relaxed.
- The reproductive parts found in the boar, stallion, dog, and buck have functions similar to those of the bull. The filiform appendage found in the ram is actually the opening of the urethra, which extends beyond the head of the penis. It rotates and sprays semen during mating. The reproductive system of the fowl, however, differs in form and function from the other domestic animals.
- Testicles – As in the other animals, sperm and hormones are produced in the testicles. In fowl, they also secrete the seminal fluid.
- Vas deferens – Sperm and seminal fluid are transferred from the testicles to the cloaca through the vas deferens.

- Cloaca – The reproductive and digestive systems meet at the cloaca. The male cloaca joins the female cloaca in the mating process.
- Papillae – Attached to the inside wall of the cloaca, the papillae transport the sperm to the female reproductive tract during mating.
- Vent – The vent is connected to the cloaca. It releases the reproductive and digestive products.

Hormones

A hormone is a chemical substance secreted by an organ of the body. When released into the bloodstream, it triggers a specific response in another organ. Reproductive hormones affect the activity of the parts of the reproductive system.

Gonadotrophin releasing hormone (GnRH) is a reproductive hormone produced by the hypothalamus gland, which is found in the brain; GnRH is not present in fowl. The anterior lobe of the pituitary gland secretes follicle stimulating hormone (FSH) as well as luteinizing hormone (LH). LH is also sometimes referred to as interstitial cell stimulating hormone (ICSH). The sex hormones that control male reproductive development and behavior are collectively called androgens. The androgen testosterone is produced in the testicles of all the species except rabbits, which produce another androgen.

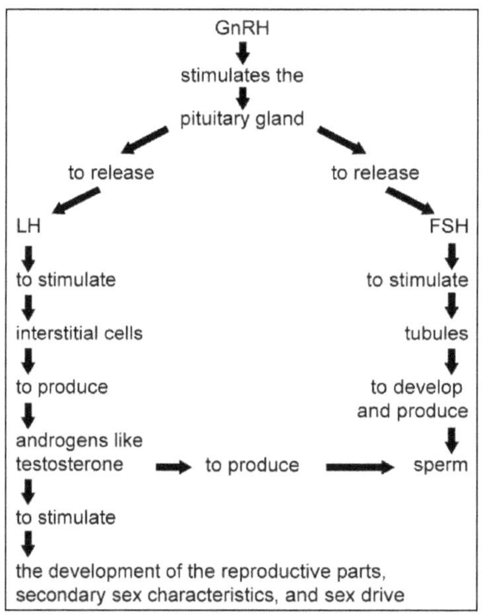

Hormones of the Male Reproductive System in Mammals.

Role of Hormones

Hormones play an important role in reproduction. A delicate balance exists among the various hormones. Figure above shows the connections between the different reproductive hormones and parts in mammals.

In mammals, GnRH stimulates the anterior pituitary gland in the brain to release both FSH and LH, which are essential to reproduction. In fowl, the release of FSH from the pituitary is stimulated by increased light. FSH affects the tubules in the testicles, causing them to produce sperm. LH causes the interstitial cells in the testicles to secrete androgens.

Androgens like testosterone stimulate the development, growth, and activity of reproductive parts. The production of androgens triggers a maturing stage called puberty, in which the reproductive parts mature as the ideal hormone level is reached. Testosterone and other androgens also trigger the development of male secondary sex characteristics. These characteristics may include a deep voice, heavy muscling, and aggressiveness; in fowl, sex characteristics may consist of plumage, a wattle and comb, and crowing. Sex drive, which is the desire and ability to mate, is similarly stimulated by androgens. They also function in the production of sperm.

Female Reproductive System

The female animal, like the male, has a reproductive system that consists of several different organs, as well as hormones that trigger the development and functioning of the various parts. The parts and hormones work together for the successful reproduction of healthy offspring.

Parts of the Reproductive System

The reproductive system of the female is very different from that of the male of the same species. However, all female mammals have a similar system. The reproductive system of the cow can therefore be used as a general example.

Cow – A cow is a female bovine. Its reproductive system is pictured in Figure below. Two ovaries are found toward the rear of the abdominal cavity. Located next to each ovary is an infundibulum, which is the funnel-shaped portion of the oviduct. The oviducts, also known as the fallopian tubes, lead from the ovary to the uterus. The uterus consists of two branches, called uterine horns, and a body. The uterus is connected to the vagina by the cervix, the inner surface of which consists of folds of tissue. The contents of the bladder are emptied into the base of the vagina by the urethra. The external parts of the female reproductive tract are the clitoris and the vulva.

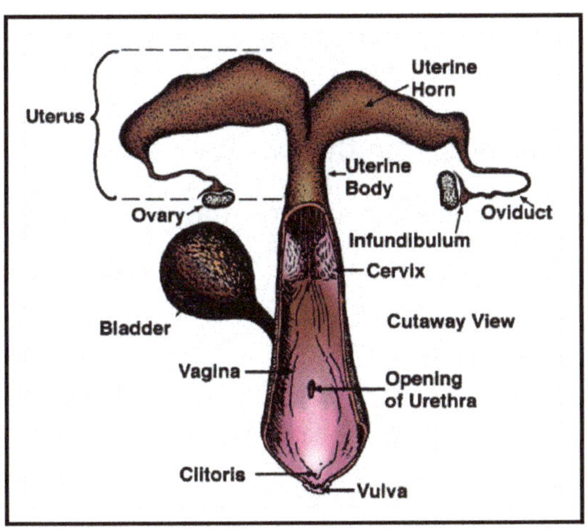

Reproductive Parts of a Cow.

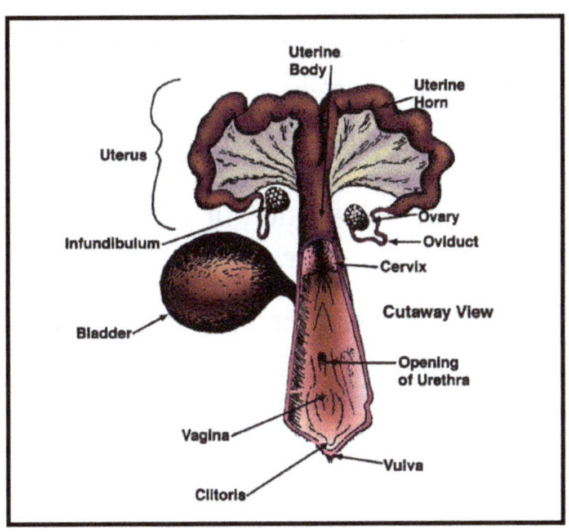

Reproductive Parts of the Sow.

Sow – A sow is a female swine. The sow's reproductive system differs from the cow in that the uterine horns are more prominent in the sow. Also, the cervix is not folded but has protruding areas instead.

Ewe – The ewe is a female sheep. Its reproductive system is very much like the cow's.

Mare – In horses, the female is called a mare. The mare has a smoother cervix, without folds like the cow's.

Fowl – The female reproductive system in fowl is quite different from that of the mammals. Figure shows the reproductive organs of the chicken. Only the left ovary and oviduct of the hen are functional, although the right is found in an underdeveloped state. The oviduct has five parts—the infundibulum, magnum, isthmus, uterus, and the vagina. A cloaca and vent are located at the end of the oviduct.

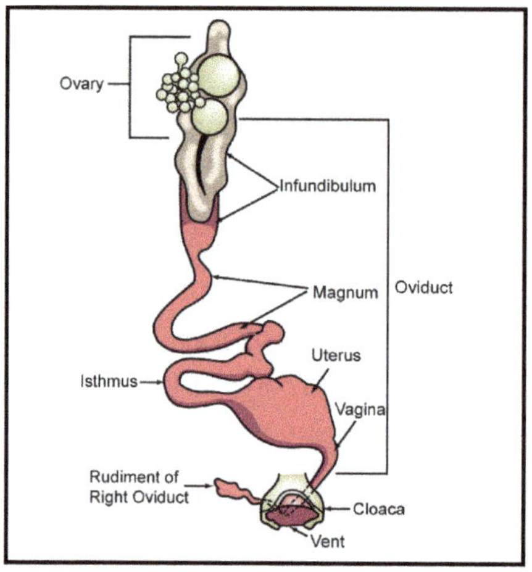

Reproductive Parts of the Hen.

Functions of the Parts

Each part of the female reproductive system has a very specific function. If one of the organs in the reproductive system does not function correctly, reproduction may be impaired. It may become difficult or impossible for the animal to produce healthy offspring.

- Ovary – The ovaries produce the female sex cells, called eggs or ova, and sex hormones. The ova are released from follicles on the ovary.
- Infundibulum – The infundibulum receives the egg from the ovary.
- Oviduct – The oviduct, or fallopian tube, is a tube-like structure that carries the egg from the ovary to the uterus. It is also the site where sperm and egg meet during fertilization.
- Uterus – The fertilized egg develops in the uterus before birth.
- Uterine horns – The fertilized egg attaches to the uterine wall and begins to develop in the uterine horns. The sow and rabbit have more pronounced uterine horns because they are litter-bearing animals.
- Cervix – The cervix is a muscular structure that functions as a passageway for the sperm and keeps bacteria and foreign material out of the uterus during pregnancy by forming a waxy mucus plug. During the birthing process, the cervix acts as a part of the birth canal.
- Vagina – Semen is deposited into the reproductive system through the vagina. The vagina is a part of the birth canal as well as part of the route by which urine passes out of the body.

- Urinary bladder – The bladder stores urine before it is released through the urethra.
- Urethra – Urine is passed from the bladder to the vagina through the urethra. It empties into the base of the vagina. The urethra and bladder have no reproductive function.
- Clitoris – The clitoris, which is located inside the vulva, is a highly sensitive part corresponding to the penis that is stimulated during mating.
- Vulva – The vulva is the external opening of the urinary and reproductive systems.
- The functions of the various parts of the reproductive system described above are the same for all these species except fowl. In fowl, the reproductive parts and their functions are significantly different.
- Ovary – As in other animals discussed in this lesson, the hen's single ovary produces ova. The nucleus of the egg is attached to a yolk sac.
- Oviduct – The oviduct is a long tube that transports the mature yolk from the ovary to the cloaca. It also carries semen to the infundibulum.
- Infundibulum – The infundibulum receives the mature yolk from the ovary. It is also the site where the egg is fertilized by the semen, which is stored in the infundibulum in the folds of the oviduct.
- Magnum – The magnum secretes albumen, which is the white of the egg. It surrounds the yolk.
- Isthmus – Two shell membranes are added to the egg yolk and white in the isthmus.
- Uterus – Also known as the shell gland, the uterus adds a thin white, shell, and pigment to the egg.
- Vagina – The egg is temporarily stored in the vagina before it is laid. It also produces the cuticle, the exterior coating of the shell.
- Cloaca – The cloaca, which is the junction of the digestive and reproductive systems, receives the male's semen. The egg passes through the cloaca during laying.
- Vent – The vent is the opening through which the egg passes out of the body during laying.

Hormones

Hormones are vital to reproduction. They cause the organs of the reproductive system to develop and prepare the body of the female for producing offspring.

Some of the hormones produced by females are also found in males. For example, as in the male, gonadotrophin releasing hormone (GnRH) is produced by the hypothalamus. Follicle stimulating hormone (FSH) and luteinizing hormone (LH) are also secreted from the anterior pituitary gland. Other hormones, however, are unique to the female, such as estrogen, which is secreted by the follicles in the ovary. Another hormone produced in the ovary is progesterone, which is secreted by the corpus luteum, a body that develops from a follicle that has released an ovum. The hormone prostaglandin is released by the uterus.

Fowl differ not only in the parts of the reproductive system but also in the number of hormones produced. Their bodies secrete only four major sex hormones. FSH and LH are released by the anterior pituitary gland, and estrogen and progesterone are secreted by the ovary. While these hormones are found in the other species, their functions differ in fowl.

Role of Hormones

Hormones regulate and develop the reproductive system. The hormones must interact in a balanced manner to ensure the proper development and functioning of the reproductive system. Figure graphically illustrates the female hormone cycle.

As it does in the male, GnRH stimulates the pituitary to release FSH and LH. In the female, FSH stimulates the follicles in the ovaries to develop a mature ovum. FSH also triggers the secretion of estrogen.

Estrogen causes the sex organs and secondary sex characteristics to develop and stimulates a desire to mate. Estrogen has an effect on other hormones as well; when estrogen is released, it suppresses the production of FSH and encourages the production of LH by the pituitary. Estrogen also causes uterine contractions that aid in transporting sperm to the oviduct to fertilize the egg.

LH causes the follicle to release the egg, a process referred to as ovulation. In addition, this hormone develops the corpus luteum, also known as the yellow body, from the follicle after the egg is released.

The hormone progesterone, secreted by the corpus luteum, stops the production of FSH and LH. It thus prevents follicle development and the secretion of estrogen.

The hormones then play different roles depending on whether the egg is fertilized or not. If the egg is fertilized, the corpus luteum stays in place, and progesterone is produced to maintain the pregnancy by preventing uterine contractions and triggering the release of secretions to nourish the fertilized egg. Progesterone also blocks ovarian activity by inhibiting the secretion of GnRH. If the egg is not fertilized, the corpus luteum deteriorates due to the secretion of prostaglandin in the uterus, ending progesterone production. When progesterone levels are low, GnRH is released and the cycle starts again.

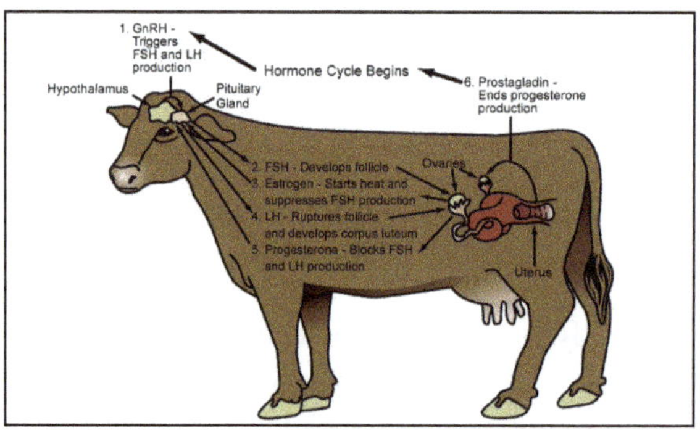

Female Hormone Cycle.

In fowl, the hormones have different roles. FSH production is triggered by increased light; for example, lengthening spring days cause FSH to be produced in wild birds. FSH's main role is to develop yolks. It also stimulates the secretion of estrogen and progesterone. Estrogen increases blood calcium, protein, fats, vitamins, and other egg formation substances. The hormone also plays an important role in egg laying by prompting the separation of the pubic bones and the enlargement of the vent. Progesterone causes the hypothalamus to trigger LH production by the pituitary. When LH is added to the bloodstream, the follicle opens and releases the mature yolk into the infundibulum.

Puberty and the Estrous Cycle

Animals must go through a process of sexual maturation before they are capable of producing offspring. It is important for the producer to be able to recognize when the animal has matured to the point that it is capable of reproduction. The producer also has to know when the female has entered a period of fertility and how long that period is likely to last. A lack of knowledge in these areas could be economically costly if productivity is lessened.

Puberty

Puberty is the stage of sexual maturation. In males, puberty is marked by a desire to mate and the ability to produce sperm capable of fertilizing an egg. Puberty in females involves the production of ova by the follicles, the development of the reproductive tract to give the animal the capacity to bear offspring, and a desire for mating.

While male animals are capable of breeding at any time after reaching puberty, female mammals are receptive to breeding only during specific periods, called estrus, or heat. Puberty in the female is marked by the first estrus. Therefore, it is important for breeding purposes to be able to recognize estrus.

Age of Puberty

Just as species differ in size and other characteristics, the age at which the female of each species reaches puberty varies. The age at puberty differs among individual animals within a particular species as well, dependent on factors such as nutrition, physical size, and breed.

In cattle, a heifer reaches puberty between six and twelve months of age; in swine, the age at which a gilt reaches puberty ranges between four to eight months. A ewe lamb becomes sexually mature between five and nine months of age. Horses reach puberty at ten to twelve months.

In contrast to these other species, fowl do not mark puberty with estrus. They do, however, pass through a period of sexual maturation in order to reproduce. For example, chickens mature sexually at 17 to 26 weeks of age. Turkeys mature later, at 27 to 30 weeks.

Estrus

Estrus, or heat, is the period during which the female will accept the male for mating. Estrus begins at puberty and occurs at intervals during a female's life. Each interval is called an estrous cycle. The purpose of the estrous cycle is to prepare the female's reproductive tract for the release of the egg and receiving the embryo.

An estrous cycle has three phases—proestrus, estrus, and diestrus. During the proestrus phase, a follicle matures to release the egg. In the estrus phase, the female is sexually receptive to the male for mating. The egg is also released by the ovary during this phase. The last phase of the estrous cycle is diestrus. During diestrus, the reproductive tract becomes less active, until the cycle begins again with proestrus.

Variation in Estrous Cycles

Each species has a unique estrous cycle. They differ in how often estrus occurs in a year or breeding season, as well as in the length of the cycle.

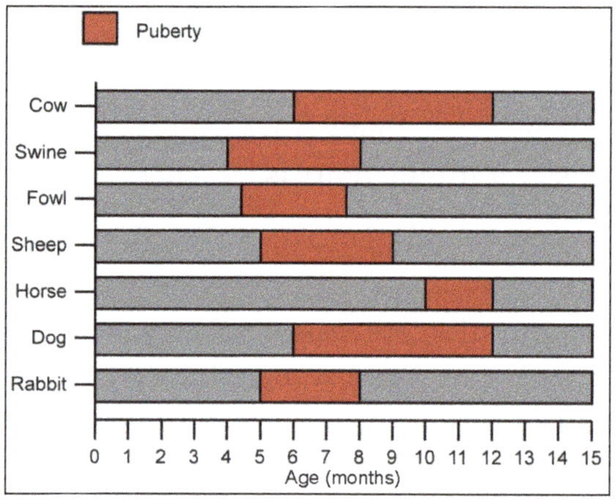

Age of Puberty in Various Species.

Animal Reproduction

A monoestrus animal has only one estrus in a breeding season or year. Polyestrous animals are in estrus more than once a breeding season or year. Cattle, swine, sheep, horses, and rabbits are polyestrous, with estrus occurring a number of times throughout the year. Sheep and horses are different from the others in that they are seasonal breeders. In most breeds of sheep, the ewe cycles only during the fall. The time when the ewe is not cycling is called anestrus. During the breeding season, the ewe is in estrus several times. The mare usually experiences anestrus in the winter and cycles from around March until October or November. The variations in the estrous cycles of sheep and horses are triggered by changes in the length of days during the year. Decreasing daylight triggers FSH production in sheep, while increasing daylight stimulates the production of FSH in horses.

Estrous cycle lengths also differ. A cow, sow, and mare all have an average estrous cycle length of 21 days, although the ranges of possible estrous lengths vary. A cow's estrous cycle ranges between 17 and 24 days, a sow's between 18 and 24 days, and a mare's between 18 and 25 days. Sheep have an average estrous cycle length of 16 days, with a range of 14 to 20 days. Rabbits have an estrous cycle of 15 days with a range of 15 to 16 days. The length of a particular animal's estrous cycle may be altered by external factors such as the animal's environment and nutrition.

Species	Estrous Cycle	Estrus
Cattle	21 days (17-24 days)	18 hours (10-30 hours)
Swine	21 days (18-24 days)	3 days (1-5 days)
Sheep	16 days (14-20 days)	30 hours (22-38 hours)
Fowl	-------	-------
Horses	21 days (18-25 days)	5 days (1-10 days)

Estrus Length

Estrus is the window of opportunity for mating. The timing of breeding is important, so a knowledge of estrus length is crucial. A cow has an average estrus length of 18 hours, with a range of 10 to 30 hours. The sow has a longer estrus period. A sow is in estrus for an average of 3 days, although the period can range between 1 to 5 days. A ewe has an estrus length of 30 hours, with a possible range of 22 to 38 hours. Mares are in estrus for an average of 5 days, with a range of 1 to 10 days.

Visual Signs of Estrus

The breeder can tell whether or not the female is in estrus by looking for visual signs, such as changes in physical appearance or behavior. These signs are triggered by the production of estrogen. The following are some of the common visual signs for each species.

Cow – A cow in estrus may mount other females or be willing to be mounted. The cow may be nervous or become agitated. Another visual sign of estrus in a cow is frequent urination. Also, it may have a swollen, inflamed vulva from which vaginal discharges or mucus are secreted.

Sow – A sow may show all of the same visual signs as a cow when in estrus. In addition, the sow may occasionally grunt loudly. It may also assume an immobile stance, holding its ears erect.

Ewe – In contrast to the cow and the sow, the ewe shows few signs of estrus. The ewe's vulva may be slightly enlarged and secrete mucus. The best indicator is the ewe seeking the ram and standing to be mounted.

Mare – Mares exhibit visual signs such as a raised tail and a relaxed vulva with mucus discharges. The mare may also disturb other mares, be nervous, nicker, and urinate frequently. A visual sign of estrus that is unique to the mare is the winking of the vulva. Winking is when the mare lowers her pelvis and raises her tail in a submissive manner, signaling that she is ready to mate.

Bitch – The bitch will exhibit changes in behavior in the days preceding estrus. For example, appetite may decrease or increase. In addition to behavioral changes, the vulva becomes swollen. A bloody discharge that begins during the proestrus period will cease during estrus. Another sign of estrus in the bitch is a willingness to accept the male for mating.

Doe – A doe in estrus exhibits a reddish-purple, slightly swollen vulva. It also accepts the male to mate.

Factors in First Time Breeding

An important factor in successful first time breeding is choosing the appropriate time for the female to be bred. Although a female can be bred at the first estrus, doing so is not recommended because the animal may not be physically equipped for successful reproduction. The three basic factors to consider when deciding whether an animal should be bred are physical size, age, and breed.

The physical size of the animal is important. The female should have the weight and frame size to carry the offspring for an entire pregnancy. Age also plays a role in determining when to breed. An animal should be past puberty and fully sexually mature. The breed of a particular animal should also be considered. One breed may develop faster than another breed within the same species and can therefore be bred sooner.

Local feed dealers, veterinarians, and extension livestock specialists are good sources of information about how to weigh these factors to determine the correct breeding age for a specific animal.

Conception and Gestation

After an animal goes through puberty, it is able to produce offspring. The process of producing a new life begins with conception and continues through gestation.

Conception

During mating, the male mammal deposits semen in the female's vagina near the cervix. The sperm travels through the cervix and the uterus into the oviduct. If an egg is present, conception occurs when the head of the single sperm penetrates the egg, which usually takes place in the upper third of the oviduct. The union of the egg and sperm creates a new life called an embryo.

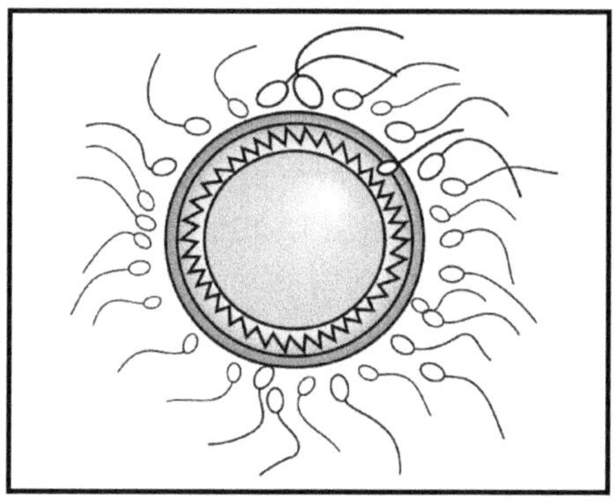

Conception

Conception is the same in cattle, swine, sheep, dogs, and rabbits. In some of these species, however, more than one egg may be released and fertilized. The sow, bitch, and doe are all litter-bearing animals. They produce multiple ova when ovulating.

While conception is similar in fowl, some differences do exist. The male deposits semen in the cloaca. The semen then travels up the oviduct to the infundibulum, where the egg is fertilized. Fertilization takes place on the yolk at a point called the germinal disk, which contains the nucleus. In contrast to the other animals, sperm from fowl remain viable for weeks while stored in the folds of the infundibulum. Conception can therefore still occur quite some time after mating.

Methods of Impregnation

A female animal can be bred naturally or artificially. Natural breeding involves the copulation of the male and the female. Artificial methods of breeding involve human intervention in impregnation.

Two methods are used to artificially impregnate animals. One method is called artificial insemination (AI). In AI, semen is collected from the male and carefully stored to ensure the survival of the sperm. The stored semen is then ready to use when the female is in estrus. The semen is inserted into the reproductive tract using an inseminating tube (mammals) or syringe (fowl).

Another method used to artificially impregnate animals is embryo transfer (ET). In ET, injections of FSH are used to stimulate the female to ovulate many eggs. The induced eggs are fertilized either naturally or artificially. Once the eggs are fertilized, the embryos are collected from the female. Individual embryos are inserted into host females for the rest of the gestation period. ET cannot be used in fowl.

Pregnancy Determination Procedures

Being able to determine whether or not an animal is pregnant is very important. It allows the producer to make appropriate production decisions for his or her animals.

One way to determine pregnancy is by a visual inspection. If pregnant, a female will not have the visual signs of estrus. Also, the female may appear to have an enlarged abdomen later in the pregnancy.

Rectal palpation is another way to determine pregnancy. It is widely used for cattle and horses. The determination is made by feeling the uterus, as illustrated in figure. After putting on a sterile glove to protect against infection, insert the hand into the female's rectum, which lies above the reproductive tract. Once the hand is inside the rectum and past the pelvic bones, feel for the cervix below. It should feel something like the neck of a soda bottle. After confirming the location of the cervix, extend the hand and arm to the uterus to feel for a fetus floating in the distended uterus. In the cow, button-like cotyledons, which are the points of attachment between the placenta and the uterus, may also be felt. A modified version of rectal palpation is used in sheep; a rod is inserted in the rectum and used to move the fetus so that it can be felt through the abdomen.

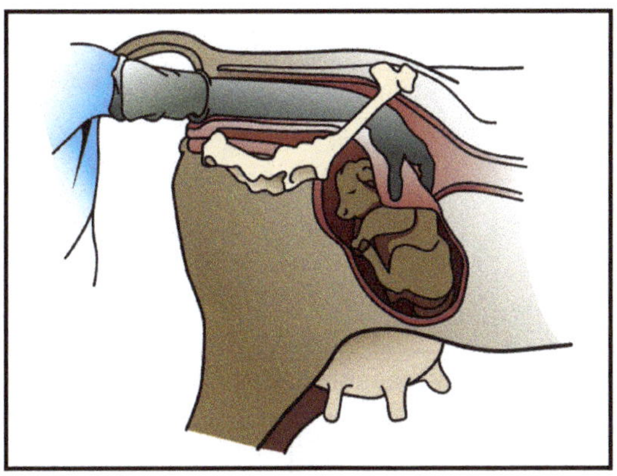

Rectal Palpation in the Cow.

Abdominal palpation can also be used to determine pregnancy. It is utilized for dogs and rabbits. Abdominal palpation is actually a routine physical check, in which the abdomen is examined externally with the hands. The person doing the palpation feels for an enlarged uterus, an indication that the female is pregnant. Care should be taken when using this method to determine pregnancy, because exerting too much pressure on the abdomen could damage the embryos.

Ultrasonic sound is used for pregnancy checks for many species. The equipment necessary includes a transducer, a machine unit, and some type of sealant. The sealant is placed between the transducer and the body to exclude air. When the transducer is correctly placed, ultrasonic sound waves are transmitted through the body. The sound waves are reflected by tissue. When the waves encounter a fetus, the unit will sound or light up, indicating that the female is pregnant. Figure shows a sow being pregnancy checked with ultrasonic sound.

Pregnancy can be determined in other ways. A laboratory may provide pregnancy results from blood tests. X-rays can be used to reveal pregnancy as well.

Gestation

Gestation is the period of development of mammalian young that begins at conception and ends at birth. Another name for gestation is pregnancy. During gestation, the fetus receives nutrients and oxygen from the mother through the placenta, the membranous tissue that surrounds the fetus in the uterus. The fetus in turn gives off carbon dioxide and waste products that are absorbed by the placenta.

Gestation can be divided into three stages, which are referred to as the cell division, embryonic, and fetal stages. The cell division stage begins right after conception. The fertilized egg undergoes 16 or more divisions while moving through the oviduct to the uterus. The cell divisions are energized by substances within the egg called cytoplasm.

The next stage is the embryonic stage. During this stage, the embryo differentiates body parts, and organs are formed. The embryo gets nourishment from the mother's uterine secretions. When the embryonic stage is complete, the fetus has been formed.

The last stage is called the fetal period. During the fetal period, the fetus continues to grow until birth. This stage is also known as the time of growth.

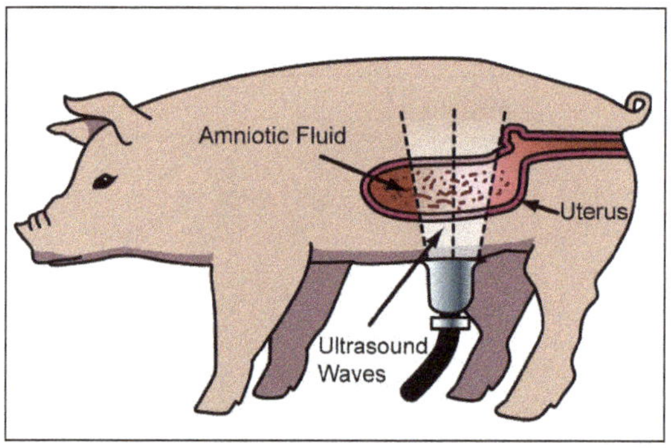

Determining Pregnancy Using Ultrasonic Sound.

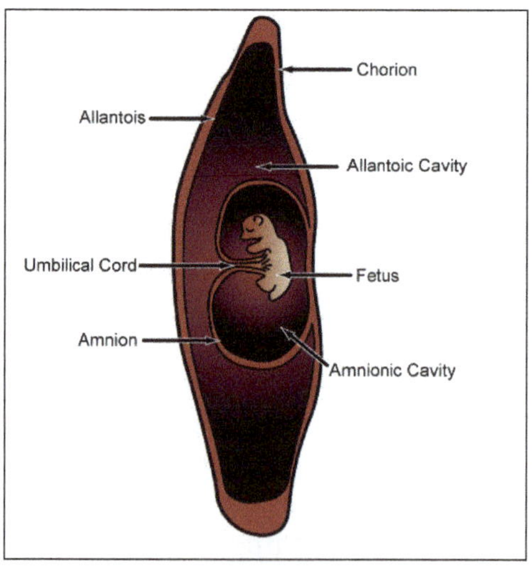

Embryonic Membranes of a Pig.

Embryonic Membranes

Three membranes form the placenta that surrounds the embryo in the uterus. These thin tissues protect and provide nourishment to the developing offspring. The embryonic membrane system is shown in figure.

The outermost embryonic membrane is the chorion, and the innermost membrane is the amnion. The chorion is connected to the uterus. Nourishment from the mother as well as waste disposal is provided through the points where the membrane is attached to the uterus. The amnion is a sac that surrounds the embryo and contains fluid. The embryo floats within the amniotic fluid, which protects the embryo from shock. The amnion is connected to the embryo's navel by the umbilical cord, which has veins and arteries that provide nourishment for the embryo.

The third membrane is between the amnion and the chorion. This membrane is called the allantois. The allantois is an extension of the urinary system of the fetus that passes through the umbilical cord. As the pregnancy progresses, the allantois occupies more space than the other two membranes. The fluid contained within the allantoic cavity comes from the kidney of the fetus.

Factors Influencing Gestation Length

The length of gestation is often influenced by the mother's characteristics. The two most important characteristics are the mother's breed and age. Gestation length may be longer for one breed than another. The mother's age can also factor into the length of gestation; an older animal may have a shorter pregnancy. In addition, gestation length is very dependent on the individual animal. Two mothers may be of the same breed and age but still have gestation periods of different lengths.

Some other factors may also influence the length of gestation. The weather, for example, may have an effect on how long gestation lasts; extremely cold weather may trigger early delivery. The sire chosen for breeding has an effect on gestation length as well. A particular sire may consistently produce offspring with a longer or shorter gestation length than the offspring of another sire.

Incubation

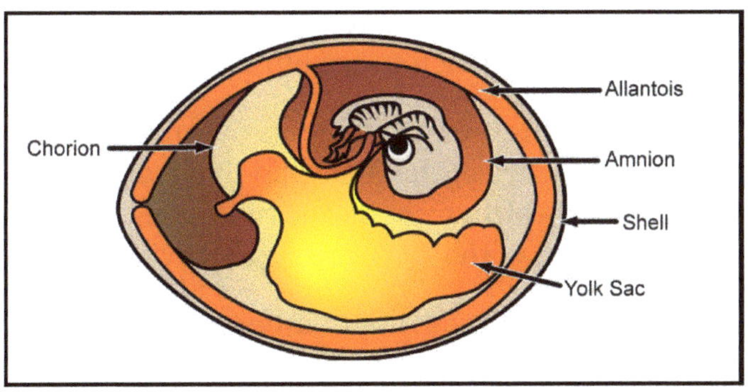

Chick Embryo.

Because fowl pass the fertilized egg out of the body surrounded by a protective shell, they do not experience a gestational period like that found in mammals. However, the embryo does go through a period of fetal growth called incubation that is similar to gestation. Incubation is the period from when a hen sits on an egg or it is placed in an incubator until the chick emerges at hatching. During incubation, the developing embryo is nourished by the yolk. The membranes that surround the embryo are the same as those found in mammals, although the chorion and allantois eventually merge to form the chorioallantoic membrane, which functions as a respiratory organ. Figure is a picture of the embryo during incubation.

Species	Gestation Length	
	Average (days)	Range (days)
Cattle	281	274-290
Swine	114	112-116
Sheep	147	144-151
Horses	336	330-350
Fowl (Incubation)	Chicken – 21	---
	Turkey - 28	---

Gestation and Incubation Lengths

Gestation lengths vary among species, as shown in table above. A cow has an average gestation length of 281 days, although it may range from 274 to 290 days. A sow averages 114 days, with a range of 112 to 116. The ewe's average gestation length is 147 days, with a range of 144 to 151 days. Gestation in a mare may range from 330 to 350 days, though the average is 336 days. A bitch may have a gestation length of anywhere from 56 to 70 days, with an average of 63 days. The average gestation length for a doe is 31 days, with a range of 30 to 32 days. The average length of incubation varies for different species of fowl, but the average is 21 days for chickens and 28 days for turkeys.

Gestation ends with parturition, or birth. Because both the mother and the new born may experience a number of complications associated with parturition, it is important for a producer to be able to recognize the signs that signal birth will soon take place, as well as problems associated with parturition. He or she will then be able to see that appropriate care is taken during and after the birth to maintain the health of both the mother and her offspring.

Parturition

Parturition is the act of giving birth. It begins after gestation has ended and is the final stage in reproduction. Hormones and mechanical factors are both at work during the birthing process.

During parturition, contractions occur to move the fetus out of the female's body. Labor involves the dilation of the cervix and the passage of the fetus and the placenta, or afterbirth. The contractions force the fetus and then the afterbirth from the uterus through the birth canal, which consists of the cervix and vagina, and then out of the body through the vulva.

Parturition is given special names when referring to different species. In cattle, for example, parturition is referred to as calving. It is called farrowing in swine, lambing in sheep, foaling in horses and kindling in rabbits.

Physical Signs of Parturition

Certain physical signs will allow the observant producer to determine when parturition will soon occur. Some signs are common to many species, while others are unique to a particular species.

Cow – An early physical sign of parturition in the cow is a swollen udder. The hips or tail head of the cow may have a sunken appearance, and the abdominal area may be enlarged or dropped. The vulva will be swollen and reddened. As birth approaches, the female will separate from the herd and locate a place to give birth. Very close to birth, the teats will swell and leak milk. Increased mucus discharge will appear from the vulva. The cow may become restless and nervously switch back and forth from a standing to a resting position. The time of birth is near when a portion of the embryonic membranes called the water bag appears from the vulva.

Sow – Signs of approaching parturition in the sow include an enlarged abdominal area, restlessness, and attempts to build a nest. A swollen vulva and teats indicate that the sow will farrow soon.

Ewe – The first physical sign of parturition that may be observed in sheep is that the teats are swollen with milk. The vulva will swell and become slack. As lambing approaches, a mucus discharge will appear from the vulva, and the ewe will become restless, changing position by standing, lying down, and then standing again.

Mare – As in the cow, early physical signs of parturition in the mare include a swollen udder and sunken hips with a dropped abdomen. The teats will swell and a waxy substance will appear in the nipples; closer to birth the wax will disappear and milk will leak from the teats. The vulva will become swollen and relaxed. The mare will seek a spot away from other horses if possible. It may exhibit a raised tail, urinate frequently, and sweat excessively. The mare will restlessly shift from a standing to a lying position. Birth should occur soon after the water bag breaks, discharging its fluid.

Hormones and Parturition

Hormones trigger parturition. Toward the end of gestation, the corpus luteum reduces its production of progesterone. As progesterone levels in the blood decrease, the levels of the hormones estrogen, oxytocin, and relaxin increase. Relaxin is produced by the corpus luteum in most species. It relaxes the pelvic muscles, cartilage, and ligaments. The birth canal then opens due to increased levels of estrogen from the ovary. The estrogen acts together with oxytocin from the pituitary to cause contractions in the uterus.

The hormone prolactin is also produced before birth by the anterior pituitary gland. It stimulates the production of milk for the new born.

Problems during Parturition

Because successful parturition depends on many factors, problems sometimes arise during the birthing process. These problems may be associated with either the mother or the fetus.

Some of the problems associated with parturition are due to the size of the fetus in relation to the mother's birth canal and pelvic opening. For example, a very large fetus may be too big to pass through the female's birth canal. A similar problem may occur with a normal-sized fetus if the female's pelvic opening is smaller than average. Problems with size arise due to the selection of animals for breeding. Parturition may be difficult if a large male of a particular breed is bred to a small female of the same breed. Similar problems occur if a male of a larger breed is bred to a female of a smaller breed.

A young mother may also have difficulty with the passage of the fetus. The lack of both birthing experience and physical development is the source of the difficulty.

During the birth, damage may be done to the reproductive system. The mother can tear the cervix, vagina, and vulva. After the birth, a prolapse may occur, with the organs of the reproductive system being pulled out of the mother's body.

Other complications may also arise. The fetus may be positioned abnormally, or malpresented, which can create birthing problems. Uncontrollable fetal bleeding can be a problem if the fetus is injured during parturition. The mother or newborn may also become infected from the actual birthing process.

Malpresentations

The fetus should be positioned in the uterus in order to pass through the birth canal with the least resistance. The normal presentation for cattle, sheep, horses, dogs, and rabbits is with the head positioned between the front legs to leave the birth canal first. Fetal pigs, however, do not orient themselves in any one position for birth.

Any variation from the normal position is called a malpresentation. For example, one malpresentation occurs when the fetus is positioned backward with both hind legs extended to pass into the birth canal first. While birth may proceed without assistance and occur without difficulty, the birth process must go quickly or the fetus may suffocate if the umbilical cord breaks.

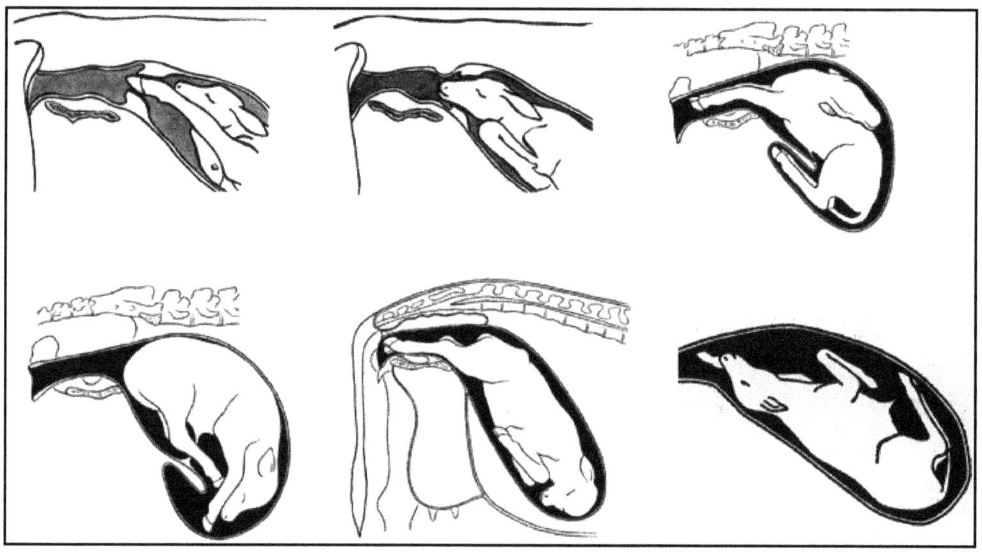

Parturition Presentations of a Calf.

Other malpresentations require intervention in order for the birth to proceed. Often the fetus can be straightened to exit the birth canal and pulled from the mother. While some malpresentations can be corrected by the producer, many require the assistance of a veterinarian. Some of the common types of malpresentations and ways to assist with each are listed below.

1. The fetus may be turned backward with its legs retained, or bent back into the uterus. If this malpresentation occurs, straighten the hind legs to enter the birth canal. The fetus must then be delivered quickly to avoid suffocation.

2. The fetus may be positioned to leave the birth canal headfirst but have one or both forefeet retained. If a foot is retained, pull it forward in order for the calf to exit the birth canal in the normal position.
3. The head of the fetus may be bent backward. To assist with the birth, reposition the fetus by pushing it back into the uterus and bringing the head forward between the legs. Pull the legs forward to allow the fetus to exit the birth canal.
4. The fetus may be presented abnormally by being turned upside down. Turn the fetus to the normal position or deliver it upside down.
5. The fetus may be turned upside down and backward. Turn it to the normal position.

Occasionally, some births are so problematic that the fetus may require additional assistance. If pulling the fetus is not appropriate, a cesarean may be performed. In this procedure, a veterinarian makes an incision in the mother's abdomen to deliver the fetus.

If the fetus is no longer alive, dismemberment of the dead fetus by a veterinarian may be an alternative to remove it from the mother. Dismemberment is done to provide the least resistance for the mother.

Postpartum Management Factors

Once the birth process has been completed, the producer still has work to do. In the postpartum period, or the period that follows immediately after parturition, both the mother and the new born need to receive care in order to ensure that they are healthy.

Cattle – Make sure that the calf is breathing properly. It may be necessary to remove mucus from the respiratory tract or provide some form of stimulation to trigger breathing. Make sure the calf has been licked dry by the mother, or dry it by rubbing it with a cloth. The newborn calf should nurse soon after birth in order to receive colostrum, a special milk secreted for only a short while after birth that passes vitamins, minerals, and antibodies to the calf; it may be necessary to hand or force feed the calf. The navel should be dipped in iodine to reduce navel infections. The calf should be identified with an ear tag. Check to make sure that the cow has expelled the afterbirth. If the placenta has been retained, serious infections can result. The cow should also be checked for signs of milk fever. Milk fever is associated with low levels of calcium in the blood and causes partial paralysis of muscles. It may appear shortly before calving and make it difficult for the cow to give birth; it may also appear after calving. Signs of milk fever are a limp tail and dragging hind feet. A veterinarian must be contacted to treat the animal with calcium injections. If untreated, the cow may become paralyzed, collapse, and die in a relatively short period of time.

Swine – Check the newborn piglet's breathing and remove mucus or provide stimulation if necessary. Dry the piglet with a cloth. Make sure that the pig nurses to receive colostrum. Remove the umbilical cord, and treat the navel with iodine. Clip the needle teeth with side cutters to protect the sow's udder. Dock the piglet's tail, and then notch the ears for identification. Make sure that the afterbirth is expelled by the sow.

Sheep – Observe the lamb's respiration. Place the lamb and its mother in a clean pen. Dry the lamb with a cloth, and make sure it is warm. It may be necessary to use heat lamps or warm water to warm the lamb, since lambs may die from becoming chilled. Check that the lamb nurses to get colostrum, and force feed if necessary. Cut the umbilical cord and treat it with iodine. Monitor the ewe to make sure the afterbirth is not retained. Check for signs of milk fever.

Horses – Check the foal's breathing. Make sure the foal nurses in order to obtain colostrum. The umbilical cord should break by itself, but if it has not broken after five minutes, it may be necessary to cut it with a pair of clean, dull scissors. However, cutting the cord may cause excessive bleeding. Once the umbilical cord has been broken or cut, dip the navel in iodine. Make sure the afterbirth is expelled within three hours; if not, call a veterinarian. The afterbirth should be closely examined to make sure the entire placenta is expelled by the mare due to the high risk of infection. Check to make sure that the foal has a bowel movement in the first 24 hours; if not, an enema may need to be administered.

Fowl and the Hatching Process

Fowl do not experience parturition, since they lay eggs instead of giving birth to young. Birds go through their own birthing process called hatching, which involves breaking through the egg's shell. In order for an egg to hatch properly, temperature, humidity, and air velocity around the egg all need to be monitored during incubation. Cool or hot temperatures, low humidity, and too low or too high a velocity of fresh air may also impair hatchability.

Like mammals, the fowl fetus may be malpresented. In fowl, the normal hatching position is with the fetus lying on its side along the longest axis of the egg. The head, which is turned to the egg's larger end, is tucked under the right wing. A malpresentation is any variation from this position.

As in the case of the young of other species, newly hatched fowl need special attention. The temperature of the brooder, which houses the young chicks, needs to be highest right after hatching. A brooder guard should be placed around the brooder for the first few days to ensure that the young birds remain where they can receive enough warmth for health and growth. Humidity and space need to be adequate, and the amount of lighting should be carefully controlled. Make sure the new fowl have proper and sufficient feed and water and are eating and drinking.

Animal Breeding

Animal breeding refers to the controlled propagation of domestic animals in order to improve desirable qualities. Humanity has been modifying domesticated animals to better suit human needs for centuries. Selective breeding involves using knowledge from several branches of science. These include genetics, statistics, reproductive physiology, computer science, and molecular genetics.

Breeding and Variation

English agriculturist Robert Bakewell was a very successful breeder of commercial livestock in the 18th century. His work was based on the traditional method of visual appraisal of the animals that he selected. Although he did not write about his methods, it is recorded that he traveled extensively by horseback and collected sheep and cattle that he considered useful. It is thought that he made wide outcrosses of diverse breeds, and then practiced inbreeding with the intent of fixing desirable characteristics in the crossbred animals. He was also the first to systematically let his animals for stud. For these reasons he is generally recognized as the first scientific breeder.

In animal breeding, a population is a group of interbreeding individuals—i.e., a breed or strain within a breed that is different in some aspects from other breeds or strains. Typically, certain animals within a breed are designated as purebred. The essential difference between purebred and non-purebred animals is that the genealogy of purebred animals has been carefully recorded, usually in a herd book, or studbook, kept by some sanctioning association. Purebred associations provide other services that are useful to their members to enhance their businesses.

Thoroughbred stallion with dark bay coat.

Selective breeding utilizes the natural variations in traits that exist among members of any population. Breeding progress requires understanding the two sources of variation: genetics and environment. For some traits there is an interaction of genetics and the environment. Differences in the animals' environment, such as amount of feed, care, and even the weather, may have an impact on their growth, reproduction, and productivity. Such variations in performance because of the environment are not transmitted to the next generation. For most traits measured in domestic animals, the environment has a larger impact on variation than do genetic differences. For example, only about 30 percent of the variation in milk production in dairy cattle can be attributed to genetic effects; the remainder of the variation is due to environmental effects. Thus, environmental factors must be considered and controlled in selecting breeding stock.

Genetic variation is necessary in order to make progress in breeding successive generations. Each gene, which is the basic unit of heredity, occupies a specific location, or locus, on a chromosome. Two or more genes may be associated with a specific locus and therefore with a specific trait. (Traits that can be observed directly, such as size, colour, shape, and so forth, make up an organism's phenotype). These genes are known as alleles. If paired alleles are the same, the organism is called homozygous for that trait; if they are different, the organism is heterozygous. Typically, one of the alleles will be expressed to the exclusion of the other allele, in which case the two alleles are referred to as dominant and recessive, respectively. However, sometimes neither dominates, in which case the two alleles are called codominant.

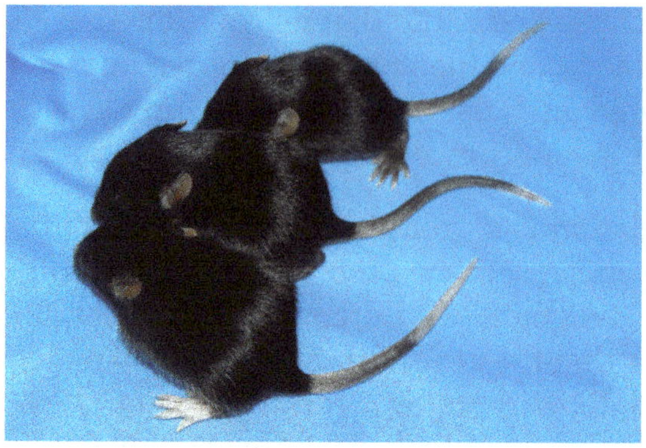

The mouse at bottom is heterozygous with a mutant gene that gives it a spotted tail. The two mice above it are paramutated; they also have spotted tails even though they do not carry the gene for this trait.

Although no complete knowledge of the genetic makeup of any breed of livestock exists yet, genetic variations can be used for improving stock. Researchers partition total genetic variation into additive, dominance, and epistatic types of gene action, which are defined in the following paragraphs. Additive variation is easiest to use in breeding because it is common and the effect of each allele at a locus just adds to the effect of other alleles at that same locus. Genetic gains made using additive genetic effects are permanent and cumulate from one generation to the next.

Although dominance variation is not more complex in theory, it is more difficult to control in practice because of how one allele masks the effect of another. For example, let a indicate a locus, with a_1 and a_2 representing two possible alleles at that location. Then a_1a_1, a_1a_2 (which is identical to a_2a_1), and a_2a_2 are the three possible genotypes. If a_1 dominates a_2, the genotypes a_1a_2 and a_1a_1 cannot be outwardly distinguished. Thus, the inability to observe differences between a_1a_2 and a_1a_1 presents a major difficulty in using dominance variance in selective breeding.

Additive and dominance variations are caused by genes at one locus. Epistatic variation is caused by the joint effects of genes at two or more loci. There has been little deliberate use of this type of genetic variation in breeding because of the complex nature of identifying and controlling the relevant genes.

Breeding Objectives

Breeding objectives can be discussed in terms of changing the genetic makeup of a population of animals, where population is defined as a recognized breed. Choice of breeding goals and design of an effective breeding program is usually not an easy task. Complicating the implementation of a breeding program is the number of generations needed to reach the initial goals. Ultimately, breeding goals are dictated by market demand; however, it is not easy to predict what consumers will want several years in advance. Sometimes the marketplace demands a different product than was defined as desirable in the original breeding objective. When this happens, breeders have to adjust their program, which results in less-efficient selection than if the new breeding goal had been used from the beginning. For example, consumers want leaner beef that is tender. Thus, ranchers have changed their cattle-breeding programs to meet this new demand. These trends have gradually changed over the last few decades; for example, Angus cattle are particularly noted for the quality of beef produced. The use of ultrasound is now widespread in determining the fat and lean content of live animals, which will hasten the changing of carcass quality to meet consumer demands.

Black Angus bull.

Additional complications arise from simultaneously trying to improve multiple traits and the difficulty of determining what part of the variation for each trait is under genetic control. In addition, some traits are genetically correlated, and this correlation may be positive or negative; that is, the traits may be complementary or antagonistic. Breeding methods depend on heritability and genetic correlations for desirable traits.

Heritability and Genetic Correlations in Breeding

Heritability is the proportion of the additive genetic variation to the total variation. Heritability is important because without genetic variation there can be no genetic change in the population. Alternatively, if heritability is high, genetic change can be quite rapid, and simple means of selection are all that is needed. Using an increasing scale from 0 to 1, a heritability of 0.75 means that 75 percent of the total variance in a trait is controlled by additive gene action. With heritability this high, just the record of a single individual's traits can easily be used to create an effective breeding program.

Some general statements can be made about heritability, keeping in mind that exceptions exist. Traits related to fertility have low heritabilities. Examples include the average number of times that a cow must be bred before she conceives and the average number of pigs in a litter. Traits related to production have intermediate heritabilities. Examples include the amount of milk a cow produces, the rates of weight gain in steers and pigs, and the number of eggs laid by chickens. So-called quality traits tend to have higher heritabilities. Examples include the amount of fat a pig has over its back and the amount of protein in a cow's milk. The magnitude of heritability is one of the primary considerations in designing breeding programs.

Female pigs can have as many as 20 piglets in a litter. China holds the record for having the largest population of domestic pigs. The United States is second.

Genetic correlation occurs when a single gene affects two traits. There may be many such genes that affect two or more traits. Genetic correlations can be positive or negative, which is indicated by assigning a number in the range from +1 to − 1, with 0 indicating no genetic correlation. A correlation of +1 means that the traits always occur together, while a correlation of − 1 means that having either trait always excludes having the other trait. Thus, the greater the displacement of the value from 0, the greater the correlation (positive or negative) between traits. The practical breeding consequence is that selection for one trait will pull along any positively correlated traits, even though there is no deliberate selection for them. For example, selecting for increased milk production also increases protein production. Another example is the selection for increased weight gain in broiler chickens, which also increases the fat content of the birds.

When traits have a negative genetic correlation, it is difficult to select simultaneously for both traits. For example, as milk production is increased in dairy cows through genetic selection, it is slightly more difficult for the high-producing cows to conceive. This negative correlation is partly due to the partitioning of the cows' nutrients between production and reproduction, with production being prioritized in early lactation. In the case of dairy cattle, milk production is on the order of 20,000 pounds per year and is increasing. This is a large metabolic demand, so nutrient demand is large to meet this need. Thus, selecting for improved fertility may result in a reduction in milk production or its rate of gain.

Methods of selection

Types of selection are individual or mass selection, within and between family selection, sibling selection, and progeny testing, with many variations. Within family selection uses the best individual from each family for breeding. Between family selection uses the whole family for selection. Mass selection uses records of only the candidates for selection. Mass selection is most effective when heritability is high and the trait is expressed early in life, in which case all that is required is observation and selection based on phenotypes. When mass selection is not appropriate, other methods of selection, which make use of relatives or progeny, can be used singularly or in combination. Modern technologies allow use of all these types of selection at the same time, which results in greater accuracy.

Elements needed to Make Genetic Progress

Genetic gain per year (ΔG) depends on balancing several factors, as expressed in the equation:

$$\Delta G = (A\sigma_g i) / I,$$

Where A is the accuracy of selection, σg is the standard deviation of the additive genetic variation in the population, i is the selection intensity (proportion selected for further breeding), and I is the generation interval (age of breeding). The σg factor cannot be easily changed within a breed, though it can be changed by crossbreeding. The other quantities in the equation can be changed. More complete pedigree records on candidates for breeding can increase the accuracy of selection, but waiting for candidates to reach full maturity in order to have better genetic data will increase the generation interval. Whether an increase in generation interval is justified by a more accurate selection process depends on individual circumstances. Selection intensity can also be increased, by narrowing the proportion of the population used in breeding, but it should be done without increasing the generation interval. Because generation interval is the divisor in the genetic gain equation, anything that increases the generation interval has an unfavourable impact on genetic progress, all else being equal.

Evaluation of Animals

Methods of ranking animals for breeding purposes have changed as statistical and genetic knowledge has increased. Along with increases in breeding knowledge, advancements in computing have enabled breeders to quickly and easily process routine breeding evaluations, as well as to develop research needed to rank large populations of animals. Evaluating and ranking candidates for selection depends on equating their performance record to a statistical model. A performance record (y) can be expressed as:

$$y = g + e + \varepsilon,$$

Where g stands for genetic effects, e indicates known (categorized) environmental effects, and ε indicates random environmental effects.

The first task in estimating g is to statistically eliminate environmental effects, a process that involves setting up a system of equations to simultaneously solve for all of the genetic effects for the sires and cows. Information from relatives is included in g and increases the accuracy of evaluation of the candidates for selection. All relatives that are available can be incorporated in this type of evaluation. This model is called the animal model.

The animal model is used extensively in evaluating beef and dairy cattle, chickens, and pigs. To apply this model for evaluating large populations requires use of high-speed computers and extensive use of advanced mathematical techniques from numerical analysis. In evaluating the dairy cattle in the United States, a system of equations with more than 25 million variables is needed.

Accuracy of Selection

In some cases the accuracy of selection for a trait can be measured using a calibrated tool or a scale. Thus, measurements of such traits can be replicated with high reliability. Alternatively, some traits are difficult to measure on an objective scale, in which case a well-designed subjective scoring method can be effective. An excellent example is hip dysplasia, a degenerative disease of the hip joints that is common in many large dog breeds. Apparently, hip dysplasia is not associated with a single allele, making its incidence very difficult to control. However, an index has been developed by radiologists that allows young dogs to be assigned a score indicating their likelihood of developing the disease as they age. In 1997 American animal geneticist E.A. Leighton reported that, in fewer than five generations of selection in a breeding experiment using these scores, the incidence of canine hip dysplasia in German shepherd dogs measured at 12 to 16 months of age had decreased from the breed average of 55 percent to 24 percent among the experimental population; in Labrador retrievers the incidence dropped from 30 to 10 percent.

Because close relatives share many genes, an examination of the relatives of a candidate for breeding can improve accuracy of selection. The more complete the genealogical record, or pedigree, the more effective the selection process. A pedigree is most useful when the heritability of a trait is relatively low, especially for traits that are expressed later in life or in only one sex.

German shepherd (Alsatian).

Reproductive techniques can be used to increase the rate of genetic progress. In particular, for species that are mostly bred by artificial insemination, the best dams can be chosen and induced to superovulate, or release multiple eggs from their ovaries. These eggs are fertilized in the uterus and then flushed out in a nonsurgical procedure that does not impair future conception of the donor female. Using this procedure, valuable females can produce more than one calf per year. Each embryo is implanted in a less-valuable host female to be carried through gestation. The sex of the embryos can be determined in utero at about 50 days of gestation. The normal gestation for Holstein-Friesian cattle is about 280 days, so this early determination of sex saves many days and allows the breeding program to be adjusted. In particular, the donor cow could be collected again, or another superior cow could be bred to produce males. Thus, these reproductive technologies reduce the generation interval and increase selection intensity by getting more than one male calf from superior females. Both superovulation and sex determination are now commonly used procedures. Superovulation is also used when breeders want to increase the number of female calves from a valuable cow.

Holstein-Friesian cow.

Progeny Testing

Progeny testing is used extensively in the beef and dairy cattle industry to aid in evaluating and selecting stock to be bred. Progeny testing is most useful when a high level of accuracy is needed for selecting a sire to be used extensively in artificial insemination. Progeny testing programs consist of choosing the best sires and dams in the population based on an animal model evaluation. A description of progeny testing in dairy breeding provides a good example. The best 1 to 2 percent of the cows from the population are chosen as bull mothers, and the best progeny-tested bulls are chosen to produce another generation of sires. The parents are mated to complement any individual deficiencies. The accuracy of evaluation of bull mothers is typically about 40 percent, and of sires that produce young bulls the accuracy is more than 80 percent. This is not as high as the industry wants for bulls to be used in artificial insemination. To reach greater accuracy, the next generation of sires is mated to enough cows in the population for each sire to produce 60 to 80 progeny. After the daughters of the young sires have a production record, the young sires are evaluated, and about the best 10 percent are used extensively to produce commercial cows.

Some of the progeny-tested sires will have thousands of daughters before a superior sire is found to replace them. About 70 percent of dairy cattle are bred by artificial insemination, so these sires control the genetic destiny of dairy cattle. Consistently applying this selection procedure has been very successful.

The genetic gain has been consistent over the years. The actual first-lactation milk production varies more than the sire breeding value because differences in environmental conditions affect first-lactation production, but these environmental effects have been adjusted out of the breeding value calculations. There is no indication that the rate of gain in the sire breeding values is about to reduce. This level of achievement can only be realized if artificial insemination organizations and producers work together.

Breeding Systems

Crossbreeding

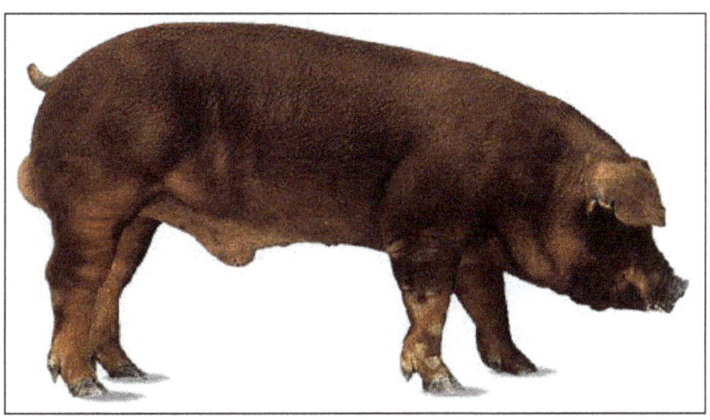

Duroc boar.

Crossbreeding involves the mating of animals from two breeds. Normally, breeds are chosen that have complementary traits that will enhance the offsprings' economic value. An example is the crossbreeding of Yorkshire and Duroc breeds of pigs. Yorkshires have acceptable rates of gain in muscle mass and produce large litters, and Durocs are very muscular and have other acceptable traits, so these breeds are complementary. Another example is Angus and Charolais beef cattle. Angus produce high-quality beef and Charolais are especially large, so crossbreeding produces an animal with acceptable quality and size.

The other consideration in crossbreeding is heterosis, or hybrid vigour, which is displayed when the offspring performance exceeds the average performance of the parent breeds. This is a common phenomenon in which increased size, growth rate, and fertility are displayed by crossbred offspring, especially when the breeds are more genetically dissimilar. Such increases generally do not increase in successive generations of crossbred stock, so purebred lines must be retained for crossbreeding and for continual improvement in the parent breeds. In general, there is more heterosis for traits with low heritability. In particular, heterosis is thought to be associated with the collective action of many genes having small effects individually but large effects cumulatively. Because of hybrid vigour, a high proportion of commercial pork and beef come from crossbred animals.

Inbreeding

Mating animals that are related causes inbreeding. Inbreeding is often described as "narrowing the genetic base" because the mating of related animals results in offspring that have more genes in common. Inbreeding is used to concentrate desirable traits. Mild inbreeding has been used in some breeds of dogs and has been extensively used in laboratory mice and rats. For example, mice have been bred to be highly sensitive to compounds that might be detrimental or useful to humans. These mice are highly inbred so that researchers can obtain the same response with replicated treatments.

Inbreeding is generally detrimental in domestic animals. Increased inbreeding is accompanied by reduced fertility, slower growth rates, greater susceptibility to disease, and higher mortality rates. As a result, producers try to avoid mating related animals. This is not always possible, though, when long-continued selection for the same traits is practiced within a small population, because parents of future generations are the best candidates from the last generation, and some inbreeding tends to accumulate. The rate of inbreeding can be reduced, but, if inbreeding depression becomes evident, some method of introducing more diverse genes will be needed. The most common method is some form of crossbreeding.

Genetic Code (DNA) and Immunogenetics

DNA

Deoxyribonucleic acid (DNA) is the genetic material that contains the instructions in each cell of organisms. DNA determines the genome, and thus the genetic code, which is a blueprint for development of all body organs and structures. The structure of DNA can be visualized as a spiral staircase. The handrails are made up of sugar and phosphate molecules, and the steps are composed of four nitrogenous bases: adenine, thymine, cytosine, and guanine. These bases are paired: Adenine is paired with thymine, and cytosine is paired with guanine. The order of these four base pairs is the genetic code that determines the genotype of an individual. The DNA is arranged on chromosomes inside cells, with cells having two methods of dividing and replicating. In mitosis, a cell divides into two daughter cells such that each contains an exact copy of the original cell's chromosomes. In meiosis, a germ cell's chromosomes are duplicated before the cell undergoes two divisions to produce four gametes, or sex cells, each with half (male or female) of the original cell's chromosomes. During the process of fertilization, male and female gametes from different organisms pair their chromosomes to form a zygote, which eventually becomes an adult.

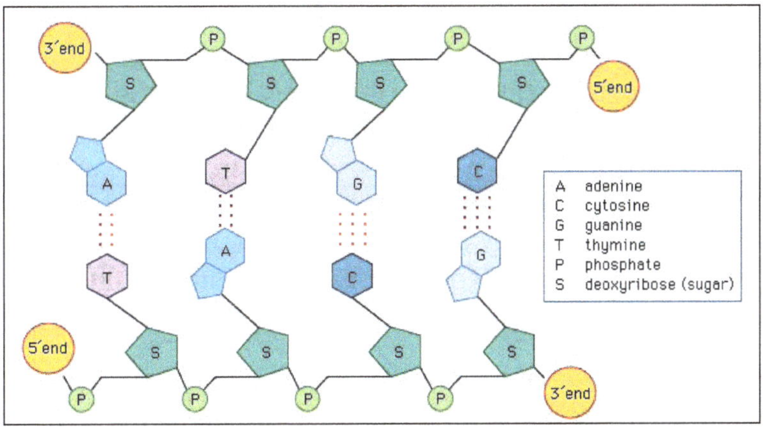

In figure, DNA structure, showing the nucleotide bases cytosine (C), thymine (T), adenine (A), and guanine (G) linked to a backbone of alternating phosphate (P) and deoxyribose sugar (S) groups. Two sugar-phosphate chains are paired through hydrogen bonds between A and T and between G and C, thus forming the twin-stranded double helix of the DNA molecule.

Genetic progress in domestic animals has been made using quantitative methods to date. It would be very desirable to know the genes that control the many traits that have economic significance in domestic animals. This should make selection more accurate. Information from sequencing human genes, as well as those of other species, is being used to find chromosomal segments with high probabilities of coding genes in livestock. Another approach is to scan a chromosome segment and look for associations with economic traits. Several quantitative trait loci have been discovered that are or promise to be useful in livestock breeding. For example, an estrogen receptor in pigs is associated with increased litter size—on average, an increase of 0.6 to 2.0 pigs per litter, depending on the genetic background in which the gene is expressed. Other genes have been found that control the secretion of casein in cow's milk. Genes are also known for growth hormone, and many others could be enumerated. With improvements in sequencing DNA, more genes will be discovered that affect economic traits—genes that will need to be tested in different genetic backgrounds and environments before they can be commercialized.

It is now much less expensive to sequence DNA, which has led to new methods of evaluating animals using large segments of 30,000–50,000 bases. With the use of these large segments of DNA, animals are evaluated without looking for markers for individual traits. This is intuitively an appealing approach because much more of the DNA can be evaluated; perhaps in the future the entire genome can be used to evaluate animals. This method of selection, called genomic selection, is now being applied to dairy cattle, but results are not yet available.

Immunogenetics

The connection between an organism's genetic makeup and its immune system, as well as applications of that knowledge, form the young science of immunogenetics. In particular, producers must control diseases in their livestock if they are going to be profitable. While vaccines, hygiene, and other therapeutic methods control most diseases, vaccines are expensive and none of these methods is completely effective. However, there is evidence from experiments and field data of some degree of genetic control over the immune system in humans and animals. For example, bovine leukocyte adhesion deficiency (BLAD) is a hereditary disease that was discovered in Holstein calves in the 1980s. The presence of the BLAD gene leads to high rates of bacterial infections, pneumonia, diarrhea, and typically death by age four months in cattle, and those that survive their youth have stunted growth and continued susceptibility to infections. It was soon found that these calves carried two copies of a recessive gene that was present in nearly 25 percent of Holstein bulls. Cattle with only one copy of the gene, or carriers, had normal growth patterns and immune systems. Holstein bulls are now routinely tested for the BLAD gene before being used for artificial insemination. With a high percentage of Holsteins being bred artificially, a potentially major problem has been avoided.

Genetic control of the immune system is based on the DNA of the individuals. Histocompatibility genes that serve several functions are on one area of a chromosome, called the major histocompatibility complex (MHC), which exists in all higher vertebrates. There are large numbers of genes involved in the MHCs of different species. There are more than 60 different alleles at one locus and other loci are multi-allelic. There are also differences among species in the number of genes known. In addition, selection experiments have demonstrated genetic variation between lines selected for high and low response to different antigens. Some vaccinations are more efficacious when the animals have been selected for resistance to the antigen for which they are vaccinated.

Substantial progress has been made in the field of immunogenetics, but limited use has been made of this knowledge. One reason for this is that immune systems have evolved to be generally robust. Changing the frequency of some genes that control immune function may inadvertently change the function of other genes and result in adverse effects. Experiments are now under way to determine whether sires' immune responses can be used to predict the health of their daughters under field conditions. The results indicate that there are differences among sires' daughter groups, but the differences are not large enough to control a high proportion of the variability. The tests used were based primarily on leukocytes, which are the first line of defense when an antigen invades an animal. Application of knowledge in the area of immunogenetics must be used with caution.

It might seem that integrating molecular markers and quantitative methods would be a trivial task. However, the effect of some genes depends on the presence of others, and these interactions need to be considered along with the particular breeding scheme. Furthermore, there are nongenetic influences that may turn genes on and off. Thus, some genes act individually, some genes interact, and the environment has a further impact. Finding how these all affect the phenotypic expression of an organism is complicated. However, this challenge presents an opportunity for future research and for producers.

Many advances in reproductive technologies have been made, though many are too expensive for everyday use. Most of the advanced techniques use artificial insemination, which was developed decades ago, though refinements continue.

Cloning

Cloning, an asexual method of reproduction, produces an individual with the same genetic material (DNA) as another individual. Probably the best-known examples of clones are identical twins, which result when cells in the early development stage separate and develop into different individuals. Though the DNA in cloned individuals is the same, environmental influences may make them differ in phenotype. Thus far, the commercial use of clones has been limited. Cloning can be used to produce clones from a highly productive individual, but the cost would have to be low enough to recover the expense quickly. Animals have been cloned by three processes: embryo splitting, blastomere dispersal, and nuclear transfer. Nuclear transfer is most common and involves enucleating an ovum, or egg, with all the genetic material removed. This material is replaced with a full set of chromosomes from a suitable donor cell, which is microinjected into the enucleated cell. Then the enucleated cell, with the transplanted chromosome, is placed into a recipient female to be carried through gestation.

Determining Sex from Sperm

There is a commercial demand for the ability to predetermine the sex of livestock. For example, a producer may want female calves from the best cows for replacements and male calves for beef production. Dairy producers may want more females for replacing cows or for expansion of their herds. The sex of mammals is determined by the sex chromosomes, or X and Y chromosomes. Animals with two X chromosomes develop into females; animals with an X and a Y chromosome develop into males. Thus, the detection of X and Y chromosomes on sperm has been the focus of research to predetermine the sex of domestic animals.

In one process, sperm is pretreated with a dye that fluoresces when exposed to short wavelength light. The fluorescence is brighter from a sperm bearing X chromosomes, which contain about 4 percent more DNA than the Y chromosome. A stream of dyed sperm is passed through a flow cytometer, a computer determines the degree of fluorescence, and the sperm is separated into different containers. The success rate can be as high as 40 percent. When "sexed sperm" has been used on a commercial basis, though, it has had limited success. The conception rate using sexed sperm is lower in cows, though it is higher in primiparous cows. In addition, sperm are killed in the typing process, and the rate of sexing the sperm is slower than desired. While economical processing of sperm is just getting started, it is expected to become another useful tool in animal agriculture.

Artificial Breeding of Animals

Artificial Insemination

Artificial insemination (AI) is the process of collecting sperm cells from a male animal and manually depositing them into the reproductive tract of a female. Artificial insemination is commonly used instead of natural mating in many species of animals because of the many benefits it can reap. These benefits include increased safety of the animals and producer, increased production efficiency and better genetics.

Artificial insemination can reduce many of the risks involved with breeding. Natural mating is a stressful process that has a much higher tendency to result in injuries or accidents of both the animals and producer. Particularly in cattle, males tend to be very large and sometimes aggressive. Artificial insemination removes all risks involved with keeping a male on the premise. Furthermore, artificial insemination reduces the risk of transmitting diseases. The entire artificial insemination procedure is much more hygienic than natural mating. All the tools and equipment are sterilized both before and after. The entire procedure is altogether a much cleaner and sanitary process.

Artificial insemination also increases efficiency. Most males usually produce enough sperm in a single ejaculate to be diluted and extended enough to create over one hundred doses. For some species, diluted semen is then packaged into "straws" and frozen. These straws of frozen semen are typically stored in a nitrogen tank where they will last for years and can be used as needed. The semen can also be shipped to various livestock producers around the United States, allowing them to artificially inseminate their herds. This gives many producers the opportunity to avoid keeping a male on the farm or potentially having to take their breeding animals to a male.

Lastly, artificial insemination can drastically improve the genetics of your herd. With artificial insemination becoming increasingly popular, many producers collect semen from their herd sires and make it available to other producers for purchase. This gives producers the opportunity to select what they want to breed their animals to.

Artificial Insemination in Cattle

In beef production systems a good reproductive performance is essential for the efficient management and production. The production of cattle can be divided into two sectors: dairy and meat production. In many European countries and developing countries, even the cattle is used as a source of meat and milk and are called dual fitness. Unlike in countries such as Australia, Brazil and the United States, the functions of production of meat and milk were separated and the creation of selective breeding is directed to a single feature.

The productive and reproductive performance of cattle herd is directly related to the service period, calving interval, number of service per conception and number of calves weaned. Artificial insemination (AI) has been proven worldwide and has proven to be a viable technical and economical to increase the genetic gain and increase efficiency, especially in production systems for meat and milk.

In cattle, AI despite presenting a series of known proven advantages, gradually being replaced by fixed-time artificial insemination (FTAI) because of lack of skilled labor, logistical problems in large AI programs, in failures detection of estrus; costs for implementing the program, no optimization of herd reproductive efficiency, and the difficulty of practical application in field conditions.

Improvement in Detection of Estrus

Worldwide there are reports that indicate low rates of service in artificially inseminated cattle, mainly due to problems in the detection of estrus. While few cows are detected in heat losses occur in significant herd reproductive efficiency, and commitment of the artificial insemination program. This commitment is even higher in Bos indicus cattle, whose breeding behavior has special features - heat of short duration with a high percentage of expression during the night. This feature was confirmed with the radiotelemetry system (Heat-watch) in Nelore, Angus x Nelore and Angus to the terrain under the same management. The results are indicative that the heat of Nelore (Bos indicus) and Nellore x Angus is about 4 hours less than the duration of estrus Angus cows (Bos taurus).

The secretion of estrogen, a manifestation of oestrus LH surge and ovulation are closely related and well known. With follicular growth, the amount of estrogen secreted increases to a peak serum concentration, triggering a preovulatory LH surge, follicular maturation and ovulation, lasting 27 hours. The goal of increased concentrations of estrogen is triggering hormonal cascade of events that includes the LH surge and a series of changes that promote follicular ovulation, and sexual behaviors associated with acceptance of mounts.

The main characteristic of estrus is the posture of immobility assumed by acceptance of the cows and ride. High producing cow's milk manifest estrus of shorter duration than cows with lower production. Females of childbearing age are pregnant or in the luteal phase of the cycle (under the domain of progesterone) are less likely to mount other females are in estrus. Almost 86% of females who ride other females are in estrus and proestrous (under the domain of the estrogen). Consequently, it should be kept open or barren females with other similarly to occur as much sexual behavioral interactions. The type of flooring is essential for the expression of mounts and immobility. When you can choose, cows in estrus spend 73% of time on the ground and not on the concrete and assemble increases by 15 times in paddocks of land in relation to the concrete. Moreover, the duration of estrus and immobility also increases on gravel.

Over the years various devices have been developed to detect estrus and they were effective as an aid in visual identification, as Bos indicus tend to present turnover night. These devices range from the tail to paint the most sophisticated pedometers and tags that track electronic activity detectors and electronic pressure HeatWatch. The pedometers have been used to measure activity or movement of the cow through a microprocessor chip and miniature device is fixed to a collar or bracelet. A cow in estrus walks about 4 times more than another that is not in estrus. These devices can be accessed manually or automatically when the cow enters the milking parlor or through receiving antennas mounted in the stalls. The information is sent to a computer and compared with individual basal activity of the same cow over a similar time interval of 2 or 3 days earlier. If the activity has increased significantly this cow, this cow is identified by a warning light that flashes on the ankle or on the computer, generating a warning report to the responsible to verify whether it is in heat and should be inseminated.

Using these devices in heifers, it was found that they are effective in identifying animals with short periods of estrus and few events of immobility. When compared the accuracy of heat identification of 49 heifers with synchronized estrus, with pedestrian detection by the farm, it failed to comply with estrus in 13 of 49 heifers (26%), while the electronic device identified all foodstuffs. Remembering that the aids can be used to increase the efficiency of our detection of estrus, but not to replace it.

Intensive detection of estrus was defined as 2 h of heat detection in the mornings and afternoons and one additional hour of estrus detection around noon. The detection was defined as a casual observation of the cows in the mornings and afternoons for 30 minutes. In the same herd, cows of both groups underwent intensive observation and casual, were synchronized with the same protocol and inseminated by the same technician. In animals with the intensive detection of estrus, the percentage of cows observed in estrus increased in 30% conception rate in 20% and pregnancy rate doubled in this group. The highest conception rates among cows intensively observed may have been the result of a more appropriate time of artificial insemination relative to ovulation.

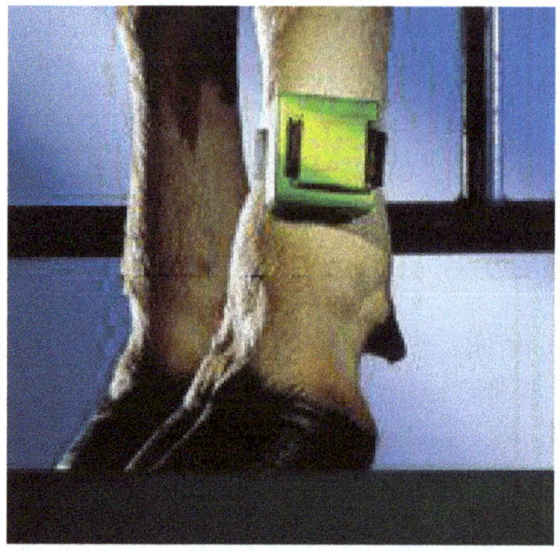

Cow with pedometer in previous member.

Whenever is observed a cow in heat, it should be removed from the herd as soon as possible because it diverts attention from other cows are in heat in two ways. First, when are making the detection of estrus, each time a cow is in a position to accept mounts noted his number. If the cow is being mounted, has been identified previously, then get distracted and fail to see other cows. Second, there is always regarded as timid cows to avoid the pushing of estrous behavior and estrous pass through undetected. These cows may show signs of estrus when the cow in estrus ruling was removed. The only time when one considers the possibility of letting the cows are in heat in the pasture to help identify other cows in heat is when you do not make use of estrus synchronization and when it is expected that less than 5% of cows in the herd come into estrus per day.

All of estrus detection devices should be used cautiously and their results interpreted by trained people, since people are the most important component in any program of estrus detection and artificial insemination. The ability to detect estrus in the interpretation and common sense are key to the success of an accurate detection of estrus and this increases the rate of submission to artificial insemination.

Reducing the Gap between Births

Calving interval (CI) covers the period between two consecutive deliveries from a cow and a yearly calving interval is a common goal for both beef producers and for the milk. Long intervals are uneconomical because cows with extended intervals produce little milk and fewer calves per year. It is one of the main parameters to measure the reproductive efficiency of cattle.

The CI can be divided into two components: the interval from calving to conception and during pregnancy. The first covers the period from calving to the establishment of a new pregnancy and is the main determinant of CI being the parameter usually manipulated to achieve a desirable CI. The second component comprises an average of 285 days, varying only according to the genetics of both the matrix and the player and can only be reduced by inducing premature parturition.

To get a CI for 365 days, the interval between delivery and conception should not be longer than 80-85 days. A key factor for achieving these goals is good nutrition. A study conducted in Florida stressed the importance of dry matter intake for cows in early cyclicity. Cows with higher dry matter intake returned to cycle faster and lost less weight postpartum. Care of the dry matter intake and body condition is essential and should begin well before birth in the last 100 days of lactation.

So there will be enough time to adjust body condition and prepare all the cows in their dry period and lactation next breeding season.

The relationship between energy balance and ovulation is directly related to the pattern of LH secretion, which is the gonadotropin responsible for follicular maturation and ovulation. After delivery, there is limited secretion of LH, but as approach the time of first ovulation, the basal concentration of LH secretion increases and the pattern becomes more pulsatile. The LH pulse frequency should occur at hourly to support follicular maturation and ovulation. The first ovulation occurs two weeks after the increased frequency of LH pulses. Thus, it should provide a well-balanced diet and palatable to cows to meet their metabolic needs as soon as possible after birth, stimulating consumption fractionated into four meals a day until the cows to produce milk and ovulate early, reducing the CI.

Another factor that must be considered is the diagnosis of pregnancy, IE this should be done up to 30 days after artificial insemination, and the empty livestock referred for procedures such as fixed-time artificial insemination. Resynchronization of repeat breeder cows is also an alternative to reduce the CI, because prior to the diagnosis of pregnancy can identify animals that are empty and reinseminated them, improving conception rates and shortening the CI. Furthermore can mention other factors that influence the duration of the CI as appropriate handling and deposition of semen, the ability of artificial insemination, animal care not to subject him to stress management, such as abuse, also contribute a good mineralization for the appearance of fertile estrus, improvement in the design and decrease the CI.

Application of Fixed-time Artificial Insemination

The fixed-time artificial insemination (FTAI) has become very popular because of reduced manpower and independence of estrus detection, since there are many difficulties to detect cows in heat, especially at night. The rates of detection of estrus and submission to artificial insemination are limiting factors due to the type of installation, which limit the expression of estrus, or lack of skilled labor for estrus detection. For artificial insemination programs are successful in herds not using hormonal intervention, the rate of estrus detection should exceed 70%. Most programs result in FTAI pregnancy rates similar to those obtained following the detection of natural estrus.

The inefficiency of reproductive cows causes great frustration and losses for farmers. Even under ideal conditions, the reproductive process is far from perfect, considering the myriad factors involved in producing a live calf. The reproductive efficiency involves not only proper management of cows, but also management of carers and inseminate. The conception rates of lactating dairy cows declined from the 50's in the USA, while the annual yield per cow increased 3.3 times. Given the inverse relationship of milk yield and fertility, there is a genetic antagonism between milk yield and reproductive traits manifested primarily in first lactation cows, but good management practices such as use of FTAI, can overcome this inverse correlation and allowed its acceptable levels of reproductive efficiency.

The programmed insemination is a method that plans and manages a program of artificial insemination of a herd. This lineup of estrous cycles has some advantages such as convenience of scheduling tasks and use of manpower, control the occurrence of estrus, ovulation, or both, and knowledge of the estrous cycle and reproductive status of females in the herd. The categories of reproductive status are scheduled to open cows at first artificial insemination, cows inseminated to be subjected to pregnancy diagnosis, cows scheduled to open re-insemination; open cows destined for disposal (will not be inseminated) and cows confirmed pregnant. Thus, the programmed insemination can be applied in two distinct groups: Cows that are scheduled for their first insemination and cows confirmed to be opened after diagnosis of pregnancy and scheduled for re-insemination. The programmed insemination involves hormonal timing of estrus, ovulation, or both. In the USA, FTAI programs are used in 90% of herds, with 86% of synchronizing first services, 77% of re-synchronization of repeaters and 59% of treatments for cystic cows in anovulatory or anestrus.

The best program is the simplest and easiest to apply, varying with the level of training of officials of the farm. In this case, simplicity means lower number of manipulations in cows during the implementation of a protocol. Employees should be trained to take a disciplined approach and comprehensive programs work, i.e., the implementation of the protocols should be approximately 100% to achieve a greater number of pregnant cows and fewer discharges of pregnant cows.

Some studies have evaluated the cost-effectiveness of fixed-time artificial insemination protocols. One study estimated the value of a pregnancy in a flock where half of the cows were inseminated only after the detection of natural estrus, compared to the application of FTAI the other half of the cows. In herds with low estrus detection, the cost of pregnancy was significantly reduced with the use of FTAI compared detection only. In herds with high rates of detection of estrus, the cost of pregnancy was higher with the use of FTAI, but there was a better reproductive performance of cows treated. The highest costs of pregnancy were associated with damage resulting from longer periods opened and account for up to 83% of total costs. Thus, the use of a system of FTAI is considered a profitable alternative for large commercial flocks in the estrus detection rates are low.

Breeding Seasons and Calving

The breeding season or insemination consists of a period of the year in which breeding females are exposed to males or are inseminated, and calving season is closely correlated, because the births are concentrated in a pre-determined by the breeding season or insemination. It is a strategy for increasing productivity in terms of number of weaned calves, zero cost, rationalizing the reproductive activity of animals with the concentration of deliveries, facilitating and regulating the management of livestock.

The most widely used method of breeding in central Brazil is one where the bull stays with the herd throughout the year. As a result, births are distributed throughout the year, despite a higher concentration during the months from July to September. The occurrence of births at times inappropriate affect the calves, due to higher incidence of diseases and parasites, or the lower availability of pasture for sows during the lactation period. The biggest drawback, however, that limits the use of rides throughout the year, concerns the difficulty of controlling the health of livestock and livestock due to lack of uniformity (age and weight) of animals. These factors ultimately affect the selection of cattle for increased reproductive potential, rather than female fertility.

The advantages in adopting the breeding season are many and are related to the calves, the matrices and the production system. With regard to calves, they will be born in more favorable time of year with a lower incidence of ectoparasites and endoparasites, with proper nutrition will have a greater supply of fodder, assist in the formation of more uniformity, reduced mortality, increased weaning weight and easy of recreating. Concerning the matrices, the mating season coincides with the increased availability of forage providing suitable conditions for the restoration of reproductive activity with higher pregnancy rates, select arrays to better reproductive efficiency after the breeding season and diagnosis gestation selecting the best females, and lactating occurring over a period of good supply of food. The advantages to the production system is characterized in rationalization of manpower, purchase of inputs less frequently, and much lower price, and ease of adoption of other practices such as early weaning, supplementation of calves, estrous synchronization and artificial insemination of matrices.

The choice of the breeding season depends on several factors such as climatic conditions, availability of pasture land, labor, adequate time for the birth of calves and purpose of production, i.e., commercial or purebred animals. Based on these facts, it is much easier to work towards the establishment of a nature to breeding season in the property, aiming to streamline the reproductive activity in both the biological and practical.

When the producer chooses to adopt the breeding season of short duration (2-3 months), replacing the system where there insemination of cows throughout the year, we recommend a gradual reduction in the period of insemination, eliminating every year one to two months to reach the optimal duration, so there is no reduction in fertility. For heifers, it is recommended to advance the breeding season in 30 days than cows, so they have more time to restore their reproductive activity when they become primiparous. Adequate protein supplementation of females during the breeding and supply of mineral throughout the year contribute to elevation of the reproduction of the herd.

Care of Semen

The canister is an insulated container with vacuum insulation, for the preservation of semen, and for that he should receive liquid nitrogen, which preserves the doses of semen frozen at a temperature of -196 °C (one hundred ninety-six degrees Celsius) indefinitely, provided they maintain a certain level, supplying them periodically. It should be handled with the utmost care to avoid damage that may result in losses. To lessen the risks to the canister, it is advisable to build a wooden box for packaging, avoiding shocks, movements are too fast, and overturn spilling its entire contents.

The liquid nitrogen evaporates constantly, and the inseminator is alert to prevent loss of semen due to lack of nitrogen. To do so, you should always measure your level with appropriate meter; never letting the level below 15 cm. High consumption of nitrogen can indicate problems with the canister as well as the formation of frost or condensate on any external surface may also indicate defects.

The thawing of semen at appropriate temperatures, according to the recommendations of the supplier is required to maximize post-thaw survival and motility of sperm. The cold shock can be avoided by maintaining the temperature, because when the blade is exposed for 30 seconds at room temperature, the temperature drops to 35 °C to 23 °C. Heating the artificial insemination pipette and maintenance of the temperature of the semen to its insertion into the vagina prevents thermal shock. Only straws should be thawed enough to inseminate cows in 10 minutes. All equipment must be kept artificial insemination extremely clean.

The deposition of semen in inadequate reproductive tract may be a limiting factor when the coach is not sure of the location of the pipette tip. Surveys show that less motile sperm reaching the oviduct when the semen is deposited in the cervix. Insemination, the goal is to reach the body of the uterus. When in doubt, it is better to deposit the semen in one or both uterine horns and fertility will be less compromised than if the semen is deposited only in the cervix. As 85 to 90% of the semen is expelled from the female reproductive tract by retrograde flow, it is essential that the total dose is deposited in the uterus.

The use of sexed semen has become common, but it is important to remember that it is different from the conventional. To achieve 90% purity of a specific sex, sperm are treated with fluorescent dyes and X and Y chromosome sperm are separated by a cell separator (flow cytometry) based on fluorescence intensity after exposure to the laser beam. There are many data Dairy Heifers, describing a design with sexed average, around 70% to 80% of the design of conventional semen used in the first service. The specific reason for this drop in fertility in artificial insemination with sexed semen, as compared to conventional, is still unknown. Nevertheless, given the potentially negative effects of the procedures for sexing, of course it is very important to the careful handling of sexed semen to optimize fertility.

Sexed semen for commercial use, is currently stored in straws thin (0.25 mL), containing 2.1 million sperm. Although 0.25 mL straws were handled similarly to 0.5 mL, the smaller diameter makes them more sensitive to errors in handling semen. The deposition of sexed semen in the uterus of the heifer must be as fast as possible, not exceeding 5 minutes. Fertility variations found after the use of sexed semen are quite large and are determined by several factors, including error handling and storage of semen. Handle carefully sexed and consider the ongoing evaluation of procedures, because every successful artificial insemination program starts with good handling practices Semen.

Body Condition Score of Females

The estimate of the nutritional status of ruminant livestock of interest by assessing body condition (BC) is a subjective measure based on the classification of animals with the coverage of the muscle and fat mass. Therefore, the body condition score (BCS) estimates the nutritional status of animals by means of visual assessment and/or tactile and represents an important tool of management. The method is fast, convenient and cheap; it reflects the energy reserves of the animals and serves as an aid in the identification of practices to be adopted in the nutritional management of the herd.

The assessment of body condition or its variation to estimate body reserves is more appropriate than the measurements of body weight, for its analysis independent of the size and physiological status of the animal. The importance of body assessment scores stems from the knowledge about the partition of nutrients according to the priority needs of the animal. The premise is to maintain life and then to preserve the species. Thus, Adams and Short proposed the following order of partition of energy nutrients: basal metabolism, mechanical activities, growth, set of basic bodily reserves of energy, ongoing maintenance of pregnancy, lactation, extra reserves of energy, estrous cyclicity, ovulation and early pregnancy, and excess reserves. Therefore, the reproductive functions, in terms of partition of nutrients, are not priorities for the animal economy.

Knowledge of body condition score herd contributes to decisions on measures of impact on production and costs of livestock development. In fact, you can set times to wean the calves or to define when and how to supplement the diet of breeders, aiming to reduce the period of postpartum anestrus. Furthermore, knowing the body condition score is useful even in the prediction of productive performance and reproductive performance.

The score is obtained by the visual and tactile (palpation) of the animal by a trained professional. There are scores of different scales, which vary in concept, the topology of the points of observation and animal species for which they are applied. The notes are given to animals in accordance with the amount of tissue reserves, especially fat and muscle in certain areas of the body, often associated with specific anatomic landmarks, such as certain bony protrusions, ribs, spinous processes of the spine, processes transverse spine, flank, tip of the ileum, above the tail, sacrum and lumbar vertebrae. Extreme scores 1 and 5 (obese and cachectic) are undesirable in any scale and in any animal species studies.

The monitoring of changes in the body condition score and body weight provides information on the reproductive potential of the cows, which is directly related to nutrition in the pre-delivery and postpartum period. Kunkle et al. found that body condition score at calving and during breeding season is closely related to the interval between births, the proportion of cows not pregnant at the end of the breeding season, milk production and cow weight at weaning. Cows with body condition at calving cycle more rapidly than those with body condition score.

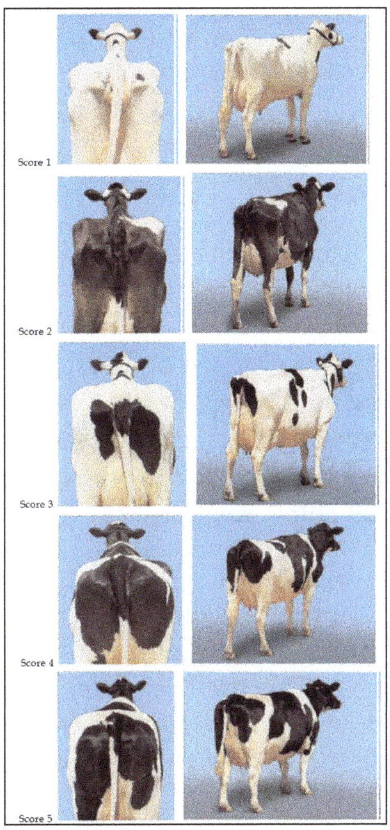

Body condition score of dairy cows.

In cows of high milk production is expected high demand for nutrients and thus mobilization of reserves in the first three to five weeks postpartum. This phenomenon is accompanied by rapid weight loss and BCS, which submits the ovarian follicles to large metabolic changes. Such variations affect the normal development of follicles and lower levels of progesterone. This scenario is associated with reduced fertility. Indeed, Walters found that the decrease in BCS after delivery decreased by 42% the quality of oocytes collected by follicular aspiration from Holstein cows.

Vizcarra et al. inferred that the nutritional status influence postpartum luteal activity and concentrations of glucose, insulin and saturated fatty acids, which are high in cows with high body condition score at calving. According to Walters, this framework explains why the delay of first ovulation in postpartum cows with negative energy balances. In fact, low plasma levels of glucose, insulin, non-esterified fatty acids and growth factor type 1 insulin are associated with inhibition of pulse frequency of luteinizing hormone and estradiol production by the dominant follicle. In cows that consume adequate dry matter during this period, follicular development is apparently normal. Already decreased by 1.0 point in the body condition score in these first five weeks postpartum resulted in lower fertility at first service. In contrast, over-conditioning, i.e. very high body condition score cows at the end of the pre-delivery caused an increase in service period and embryonic mortality.

The productivity and profitability of farms are closely related to achieving high reproductive rates, which are only achieved through the adoption of certain management practices. Among these, science-based nutrition should provide the matrix of metabolic conditions ideal to meet certain strategic moments of the production cycle, such as the breeding season, the season of birth and lactation season. In this context, the body condition score is a useful tool in assessing the nutritional status of the animal and therefore has strategic application in reproductive management of herds that are artificially inseminated.

Evaluation of Female Gynecology

All females of reproductive age in a herd must be submitted to gynecological examination for selection of suitable animals for artificial insemination program. This is an internal examination by rectal palpation, ultrasound and vaginoscopy and can be complemented by laparoscopy and biopsy. On rectal palpation and ultrasonography are checked the size, consistency and contraction of the uterus, uterine horns and symmetry. In the ovaries are observed consistent form and size of follicles, cysts and persistent corpus luteum. Vaginoscopy complements rectal palpation and ultrasonography, because it turns out the shape of the vaginal portion of cervix, the opening degree of the cervical canal, mucosa color, moisture content and characteristic vaginal and cervical mucus.

The pelvic examination should be thorough, seeking to know and understand the best possible animal's reproductive status, either for a simple confirmation of pregnancy or to identify diseases or reflection of management that is harming their reproductive efficiency. Gynecological examination involves a complete evaluation of all components of the external and internal genitalia, with emphasis on the ovaries, combining the findings of the examination with a score of animal body, with its history and with the herd.

For purposes of gynecological evaluation should consider the following groups of animals: from 20 to 30 days postpartum, postpartum with abnormal vaginal discharge, irregular estrous cycles, not seen in estrus 60 days postpartum; covered or inseminated by two or more times, and that return to estrus 45 days after artificial insemination. In the historical survey of the animal should always consider the age, number of childbirths and their conditions and cyclicality observed, as these factors may reflect the ovarian function and reproductive efficiency. Are essential, too, about the coverage, treatment for retained placenta and uterine infection, drugs and dosages, treatment outcome, nutritional program for the animals used in pre-natal and post-partum, body condition at calving, stage lactation, milk production and herd health program.

It is crucial at the end of gynecological examination, classify the animals examined according to their reproductive status in pregnant and not pregnant, the latest being identified as normal or with any individual or reproductive problems reflecting a management problem. The results of gynecological examination of different groups of animals held in a coherent and detailed by a veterinarian, provides a satisfactory understanding of the breeding herd during the exam, allowing them to be defined and adopted measures to keep the reproductive efficiency of flock.

Diagnosis of Pregnancy

Pregnancy diagnosis is an important tool for the management of rural property. This must be done by a veterinarian trained at around 28-30 days after artificial insemination by rectal palpation and/or ultrasonography. There are other methods that can be used for diagnosis of pregnancy such as: no return to service, measurement of progesterone, pregnancy-specific proteins, estrone sulfate, breast enlargement and abdominal distension. The accurate diagnosis of pregnancy is important in establishing and maintaining optimal reproductive performance. The producer should know as early as possible if the female is covered or not pregnant for it to be inseminated again.

A cow is diagnosed not pregnant if she has been observed in estrus approximately 21 days after artificial insemination. The percentage of cows not observed in estrus around this period is known as the rate of no return and is not a method to estimate pregnancy rates. As the percentage of cows ovulating truly seen in estrus is often low, the rate of return not overestimates the rates of pregnancy. If a cow is in estrus detected three weeks after artificial insemination, it can still be pregnant and her artificial insemination can cause miscarriage. One must be careful to confirm that it is true and not in heat pregnant through other methods of pregnancy diagnosis.

The detection of pregnancy by measurement of hormones, especially if they occur in milk has advantages when interference is minimal with the cow and risk-free pregnancy. The measurement of progesterone to verify pregnancy also offers the possibility of diagnosing the twenty-first day. In pregnant cows, progesterone concentrations in milk and blood remain high between the twenty-first and twenty-fourth day after ovulation, when they would be basal in the animal not pregnant. The simplest procedure is to obtain a sample of milk from the cow 21 days after artificial insemination. If the progesterone level is low, the cow is not pregnant. If the progesterone level is high, the cow may be pregnant.

The factor of early pregnancy is a gestation dependent protein complex that has been detected in the serum of various species. It is detected using an immunological technique, the rosette inhibition test. It is expected that this substance appear in the serum soon after conception and disappear very quickly after embryonic death. Two additional proteins, protein B pregnancy specific bovine and bovine pregnancy associated glycoprotein, can also be measured during early pregnancy in the cow.

The availability of ultrasound has been a major advance in the diagnosis and monitoring of pregnancy reproductive taking the advantage of not being invasive. The baby begins to fill the uterine horn near the seventeenth day of gestation and can be seen as a non-echogenic area. In the nineteenth day, the amniotic sac has expanded considerably and the lumen of the uterus can be observed. In the twenty-second day you can hear the heartbeat with the thirtieth day embryo and fetus is very visible.

Rectal palpation of the fetus relies on the ability to detect the presence of a fetus growing in one of the uterine horns by inserting the arm into the rectum of the cow. This can be a dangerous procedure, since trauma can be generated in both the cow and the fetus and must therefore be done by a trained examiner. In non-pregnant animal and the animal in early pregnancy, the uterine horns can be felt in size and approximately equal diameters. It is possible to detect a difference in the size of the two uterine horns from 35-40 days of gestation forward.

The identification of components produced by the fetus, rather than the mother, has advantages in the diagnosis of pregnancy. Estrogens are produced by the bovine fetus and the concentrations of estrone sulfate in maternal plasma increases from the seventeenth day of gestation. The content of estrone sulfate in milk reflects the plasma and estrone sulfate test positive at 15 weeks of gestation provides 100% effective diagnosis of pregnancy. In most cases, detection of non-pregnancy at this late stage is of little value in reproductive management, but can be used as a confirmatory test after diagnosis of early pregnancy.

The mammogensis or mammary gland development as a result of pregnancy can be detected in heifers as early as four months of pregnancy. Promoters like growth steroids can stimulate similar changes which may be a confounding factor. It is only during the last days of gestation, when the udder is distended with colostrum, the breast development may be regarded as accurate diagnosis of pregnancy.

The abdomen of the pregnant animal is getting distended around the seventh month of pregnancy. If a hand is pushed firmly against the right side of the abdomen, the fetus can sometimes be felt rebounding against the hand. It is not a reliable indicator of pregnancy. Pregnancy diagnosis is a vital aid for reproductive management and must be sought the best combination of earliness and accuracy.

Calving Rate

Cows get pregnant has always been and will remain the major challenge of cattle. Over the years many methods evaluation were developed, but, unfortunately, with the interval between data calving, days in milk at first AI, percentage of pregnant cows on visit the veterinarian and the first AI conception do not tell the whole story, that is, do not reflect reality.

Averages can leave much to be desired, especially when analyzing calving interval. Two herds may have intervals of 13 months births, but in the herd A the cows were pregnant in the early part and the remaining lactation and late lactation and few between. While in herd B pregnancies are distributed during lactation, and the vast majority of cows becomes pregnant at the beginning of lactation and the rest evenly distributed during lactation. Herd B has a performance reproductive better, but the average interval between births of the two herds is equal. The most reliable measure that reflects what is happening in the herd and has resulted in the birth of calves is the calving rate.

The calving rate is the percentage of cows calved in the total of pregnant cows at the beginning of the breeding season. Even under ideal conditions with 100% of normal cows and 100% efficiency in detection of estrus, farrowing rates will fail to reach 100%. Only 60-70% of inseminated cows produce a calf born alive and the great majority of failures occur before the second half of pregnancy. This is partly due to the failure of design and partly of embryonic or fetal death. The proportions of embryonic or fetal death are far greater than the failures in the design and the vast majority of these occurred probably by genetic abnormalities in embryos, but this hypothesis has never been proven. The cause is probably multifactorial, involving interactions between genetics, environment and management.

Artificial Insemination in Pigs

Artificial insemination (AI) of swine is widely practiced in countries with intensive pig production. In Western Europe, more than 90% of the sows have been bred by AI for more than two decades. When compared with natural mating, AI is a very useful tool to introduce superior genes into sow herds, with a minimal risk of disease. The outcome of AI largely depends on the semen quality and the insemination procedure. In practice, fresh diluted semen for intra-cervical insemination is mostly used in pigs. Semen is obtained from boars on farms or from specialised AI-centres. The latter offer a diversity of breeds and genetic lines and distribute ready-to-use semen doses of constant quality to different sow herds.

Three important aspects should be considered. Firstly, only semen from healthy boars should be used, as diseased boars may ejaculate semen that is contaminated with pathogens. The semen from commercial AI-centres is shipped to a large number of sow farms. Contaminated semen could therefore lead to a rapid transmission of pathogens and to disease outbreaks in many different sow herds. Strict regulations and guidelines to prevent disease spreading are therefore implemented on porcine AI-centres. The second important aspect is the fertilizing capacity of the produced semen doses. The fertilizing potential of a semen dose is inherently linked to the quality of the spermatozoa itself. Examination of the ejaculates is therefore necessary. A third important aspect of AI-centres is the semen processing procedure. This is not only important to guarantee a low microbial presence but even more to obtain high quality sperm, namely viable spermatozoa in ready-to-use semen doses that can be used for several days. The dilution procedure and semen handling, the properties of the extender and the micro-environment for the sperm cells influence the survival and longevity of the spermatozoa.

Collection of Semen, Dilution and Processing

Although automated semen collection systems have been developed, semen is mostly collected by the gloved hand technique from a boar trained to mount a dummy sow. Dummy sows should be solid in construction without sharp edges, and located in a quiet designated semen collection room with a non-slippery floor. A pre-warmed (38 °C) collection container is used. The top of the container is covered with cheesecloth to filter out gel portion of the semen. The end of the penis is grabbed firmly with a gloved hand and the collection process is initiated with firm pressure to the spiral end of penis with the hand so that the penis cannot rotate. This process imitates the pressure applied by the corkscrew shape of the sow's vagina. Polyvinyl gloves can be used, not latex gloves as these are toxic for the semen. The first part of the ejaculate (pre-sperm) should be discarded. It is clear, watery fluid and does not contain sperm (~25 ml), but it may have a high bacterial count.

The sperm-rich fraction should be collected (40-100 ml). It is very chalky in appearance and contains 80-90% of all sperm cells in the ejaculate. Once the sperm-rich fraction is complete, the remainder of the ejaculate is again clearer, watery fluid, and should not be collected (70-300 ml). After collection, the filter with gel should be discarded, and the collection container should be placed in warm water. The semen should be extended within 15 min. after collection. The ejaculation lasts up to 5 to 8 min, but may continue up to 15 min. About 100 to 300 ml of semen is collected. Semen collection from boars in AI-centres is performed approximately 2 times per week.

The extension process should be done in a warm room with clean and sterile equipment. The extender is added to the semen, and cold shock should be avoided by diminishing the temperature gradually. A normal ejaculate usually contains enough sperm to inseminate 15 to 25 sows using conventional AI. Each dose should contain 2-3 billion spermatozoa in 80-100 ml.

Semen Quality Assessment

Assessment of the Concentration of Spermatozoa in the Ejaculate

The number of spermatozoa in a semen dose is important for the fertilization process. On the other hand, AI-centres tend to dilute the ejaculates as much as possible to maximize semen dose production. Variation in the number of spermatozoa in an ejaculate has been described between different pig breeds e.g. Landrace, Duroc and Yorkshire, which is a first factor influencing semen dose production. Not only differences in sperm number but also in sperm volume, ranging from 100 to 300 ml, influence sperm concentration. Individual variation within a breed is also very important. Xu et al. demonstrated a difference in litter size of 0.09 to 1.88 piglets when inseminating sows either with 2×10^9 or 3×10^9 spermatozoa. Differences in litter size between both semen doses were largely dependent on individual variations between boars. Another study using 2×10^9 spermatozoa per dose mentioned not only a smaller litter size but also a lower farrowing rate at lower semen dose. In addition, several studies described lower fertility results when lower semen doses were used in boars with suboptimal semen quality. A general guideline for the number of good quality spermatozoa in a semen dose was set at 3×10^9 spermatozoa per dose. According to the morphological or motility characteristics, the number of spermatozoa should be adapted. By multiplying the total volume of the gel-free ejaculate (ml) times the sperm concentration per ml, the total sperm number is calculated. The volume is routinely measured by weighing the ejaculate considering 1 gram equal to 1 ml and the obtained total sperm numbers is a good indicator to evaluate spermatogenesis. The data above clearly indicate the importance of an accurate determination of the concentration of spermatozoa in the ejaculate.

Inspection of the Raw Ejaculate

Visual evaluation of the opacity of the ejaculate gives a rough idea on the sperm concentration. However, this method is crude and very subjective and therefore not suitable for AI-centres with large semen production.

Counting Chambers

Different glass chambers are described to count cells in a known volume. Haemocytometers, such as the Neubauer, Thoma and Bürker chamber are reusable glass chambers with fixed volume used for counting immobilized spermatozoa in a grit. Other reusable glass chambers as the Mackler chamber are used for assessing concentration as well as motility. Disposable chambers (Microcell^R, Leja^R) are commonly used in Computer Assisted Semen Analysis (CASA) since their small depth limits movement in the third dimension (Z-axis) when the sperm path is analysed. Haemocytometers are considered as the standard method for determining sperm concentration and have a lower coefficient of variation than disposable chambers. The concentration determined by the haemotocytometer however, was generally higher than the concentration determined using other chambers, especially with increasing concentration.

Makler chambers were described as having higher standard deviations and more inconsistent results compared with the haemocytometer. Disposable chambers on the other hand, although they are also used for counting live cells, were reported to be more consistent and accurate. The accuracy of different counting chambers is also dependent on the manner in which the chamber is filled. Thin, capillary-filled, disposable chambers are generally found to underestimate sperm concentration due to the Segre-Silberberg effect. However, the variations between chambers when analysing sperm concentration seems to be technician and laboratory dependent.

Photometry

Photometers (single wavelength) or spectrophotometers (multiple wavelengths) measure the optical density, i.e. the relative absorption and scattering of a light beam that is sent through a semen sample. The absorption and scattering is proportional to the sperm concentration. Next to the concentration of spermatozoa, the absorbance is also influenced by gel particles in the seminal plasma or the extender, by the quality of the sample cuvette, and the dilution of the sample. Photometry is commonly used in practice because it is fast and easy to perform. Accurate dilution and a correct calibration curve are imperative to obtain proper results.

Flow Cytometry

Several studies use fluorescent dyes that stain intact or damaged spermatozoa differently, and measure the distribution of dyes in the sperm cell population by a flow cytometer. In that way, viability as determined by the percentage of intact cells as well as the concentration of spermatozoa in an ejaculate can be determined by using fluorescent microspheres. Since this technology can discriminate interference from gel particles, it has a low coefficient of variation (3.3%). However, the high costs and the dependence on qualified personnel make flow cytometry not the most suitable method for use in practice.

Other Methods

Coulter counters, determining the number of particles within a known volume, can be used to assess spermatozoa concentration but discrimination of other particles with comparable size within the sample is difficult, resulting in a lower accuracy. Other systems, e.g. computer assisted semen analysis (CASA) use image analysis to determine sperm concentration within counting chambers. The accuracy of these systems depends not only on the optical properties and the software settings, but also on the kind of counting chamber that is used. Nucleocounters are also used for determining sperm concentration, and they provide similar counts as those obtained with photometers. In these devices, DNA is fluorescently labelled and counted by image analysis resulting in an accurate determination of sperm concentration.

The concern to obtain a correct estimate of sperm concentration led to a discussion on the accuracy of the different systems. Maes et al. did not find major differences between two types of colorimeters, the Bürker counting chamber, and the Hamilton Thorne Analyzer (Ceros 12.1) using two Leja chambers. Every system has its advantages and limitations. They concluded that in commercial porcine AI-centres, economic considerations such as purchase prices, labour, and high sample throughput are also important in the choice for one method or the other.

Morphology and Viability Assessment

The microscopic appearance of spermatozoa can give information on morphological abnormalities, cell membrane integrity and the acrosome. These are three important parameters that contribute to the fertilizing capacity of the sperm cells. Morphological abnormalities give an indication of aberrations in the spermatogenesis. Some malformations compromise the function of the cells and cannot be compensated for, therefore leading to culling of the boar. Abnormal shape of the head which carries the genetic material or abnormalities of the mitochondrial sheet which is important for the function of the flagella, are therefore called primary abnormalities. Remainders of cytoplasm, proximal or distal droplets, and small tail abnormalities are called secondary abnormalities and can be compensated for by the semen dose. Additionally, morphological anomalies (e.g. coiled tails) acquired by inappropriate handling of semen are called tertiary abnormalities.

Morphology can be assessed by staining techniques that do not require highly qualified personnel. Normal morphology is correlated with fertility, and should therefore be performed routinely in porcine AI-centres. Criteria for the maximum percentage of primary and secondary abnormalities in commercial porcine AI-centres were determined as 10% and 20%, respectively. The percentage of spermatozoa with normal morphology should be at least 70%. An overview of the criteria for use of porcine semen in artificial insemination is shown in table below.

Semen parameter	Requirement (%)			
	Kuster and Althouse, 1999	Martin-Rillo et al., 1996	Shipley, 1999	Britt et al., 1999
Motility	“/ 70	60-100	“/ 70	“/60
Abnormal morphology	< 20		< 20	
Normal acrosomes		<10		
Cytoplasmic droplets	< 15		< 15	
Proximal droplets		0-20		<20
Distal droplets		0-30		
Coiled tails		<5		
Primary abnormalities				<10
Secondary abnormalities				<20

Recommendation for 2×10^9 spermatozoa per dose. Overview of the cut-off values for porcine semen quality in artificial insemination.

Although several stains can be used, staining spermatozoa of farm animals for morphological examination is usually combined with membrane integrity assessment using a dye that is excluded by live cells, such as eosin. Therefore, besides being helpful for assessing sperm morphology, the eosin-nigrosin stain can be used to discriminate between live and damaged cells. This staining technique is widely used and is considered a simple and reliable technique that is easy to applyand its outcome correlates with fertility.

Next to visual morphology assessment, automated CASA-systems were developed to obtain more objective information. Automated Sperm Morphology Analysis systems (ASMA) are able to locate the head of the spermatozoa and compare its morphology to internal standards. A disadvantage is the lengthy analysis which is required for each sample, and which is partly dependent on the contrast of the cells from the background necessary for the system to recognise the cell. The prolonged analysis time undoes the advantage of an objective measurement and renders the method not suitable for use in commercial AI-centres at the moment.

Several fluorescent dyes can also be used to assess cell membrane integrity and can be combined with flow cytometric analysis. The need for qualified personnel and a fluorescence microscope excludes the use of the latter stains from commercial AI-centres, although attempts have been made to incorporate this technology in the most recent generation of CASA-instruments. Next to staining methods, membrane integrity can also be evaluated by testing the osmotic resistance of the cells. The osmotic resistance of the porcine sperm cells was correlated with fertility results. More advanced methods measure cell volume by detecting voltage changes when cells pass a capillary pore in a CASY 1 cell counter. Osmotic reactivity is a sensitive detection method of changes in plasma membrane, either in damaged cells or in capacitating cells.

Since the acrosome is important for the penetration of the oocyte, its integrity is considered vital for an optimal fertilising capacity. The acrosome integrity can be evaluated on the basis of its microscopic appearance, either by phase contrast evaluation of glutaraldehyde fixed spermatozoa or by various staining methods.

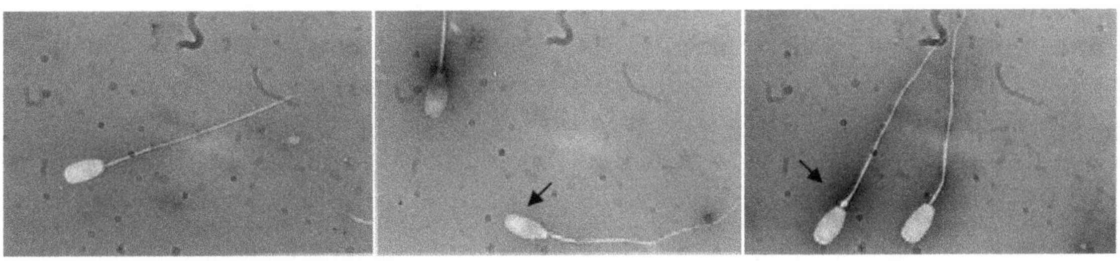

Sperm morphology: Spermatozoa with normal morphology, abnormal (narrow) head (primary defect) and proximal droplet (secondary defect) (arrows).

Motility Assessment

Motility of spermatozoa has always been considered a primary requirement to fertilize eggs. Although the spermatozoa are brought to the fertilization site mainly by uterine contractions, sperm motility is required for penetration of the zona pellucida. Motility is known to be an important characteristic in predicting the fertilizing potential of an ejaculate. Therefore, several methods have been used for motility assessment.

Visual Motility Estimation

The simplest way to evaluate sperm motility is by estimating the number of motile spermatozoa under a light microscope or using phase contrast microscopy. This method is subjective since it depends on the interpretation by an individual. It is however a cheap method and facilitates a high sample throughput which makes it popular in commercial AI-centres.

Computer Assisted Semen Analysis (CASA)

Using digital image analysis, sperm cell tracks are analysed in different components. CASA has major advantages: the method is objective, independent of the interpretation of the technician and gives detailed information on the sperm movement. This way, different motility patterns can be observed, e.g. progressive movement versus hyperactivity and even different subpopulations of spermatozoa within an ejaculate can be demonstrated. The detailed information given by the CASA-systems renders them also very susceptible to external influences on sperm movement, such as operator variability, semen handling and system settings are causes of inter-laboratory differences. At the moment, CASA instruments have been validated for many animal species which makes the method available for use in veterinary practice or commercial AI-centres. The high cost of the equipment compared to the alternative visual motility determination, is a restraint to the use of CASA in practice.

Sperm Quality Analyzer (SQA)

The SQA systems convert variations in optical density into electrical signals to determine sperm concentration and motility. These electronic signals are analyzed by the SQA software algorithms and translated into sperm quality parameters. The effectiveness of different SQA systems for sperm analysis has been studied both in humans and animals, and different algorithms are needed for each species. A previous version of the SQA namely the SQA-IIC was consistent and suitable for the estimation of boar semen quality. There was a good correlation between the sperm motility index (SMI) obtained by SQA-IIC and several CASA parameters, especially with the percentage of motile sperm and with straight line velocity (VSL). However, the SQA-IIC is based on an old technology meant for human sperm analysis and the SMI values are based on overall information of the quality of the sperm, and do not discriminate between concentration, morphology and motility parameters. Recently, the SQA-Vp was introduced as an SQA device specifically designed for boars in which the sperm movement can be visualized on a screen and motility is given as percentage of motile sperm.

In pigs, a motility score of 60% motile cells, independent of the method of assessment, is required to be considered as a fertile ejaculate. Above 60% motile spermatozoa, no differences in farrowing rate and litter size were recorded. Apart from morphology, several attempts were made to correlate motility with fertility outcome. When using adequate numbers of spermatozoa per insemination dose (3×10^9), correlation with fertility outcome was hard to establish. At lower semen dose, motility was well-correlated with fertility parameters. In most studies involving pigs, the predictive effect of motility was evaluated using visual motility assessment. To increase the discriminating power of the motility estimation, objective motility assessment by CASA-measurements or motility of spermatozoa subjected to a percoll gradient were used. These studies found motility to be positively correlated with fertility, especially with litter size.

Other Sperm Examination Techniques

Sperm examination techniques requiring specialized knowledge and expensive equipment are not frequently used in commercial AI-centres. They are however used for research on porcine sperm. DNA fragmentation tests can be used to identify subfertile boars, but the study results are contradictory. Some metabolic responses of sperm like resistance to oxidative stress and in vitro fertilisation assays are also used for research purposes. The practical relevance of these techniques is limited due to the fact that most of the research on porcine semen is based on the semen from good performing boars. Subfertile or nonfertile boars are rapidly culled because of economic considerations, and therefore, there is a lack of information regarding sperm quality of infertile boars.

Storage of Liquid Semen

Frozen storage of boar semen still yields inferior fertility due to the loss of membrane integrity during freezing and thawing. Consequently, freshly diluted semen (liquid semen) is widely used for AI on the day of collection or in the following days. For storage of liquid boar semen, two factors are very important: The temperature of collection and storage, and the composition of the storage medium.

Temperature of Collection, Transport and Storage

A different composition of the phospholipids in the membrane of boar spermatozoa compared to bull spermatozoa, a low cholesterol/phospholipid ratio and an asymmetrical distribution of cholesterol within the membrane render boar spermatozoa very susceptible to cold temperatures resulting in increased permeability and loss of controlled membrane processes. Hence, rapid cooling of ejaculates to 15 °C or cooling below 15 °C results in loss of viability or cold shock. To avoid this cold shock, prediluted ejaculates are better left at temperatures above 15 °C for several hours to induce cold resistance. In practice, semen is collected in isolated cans to avoid contact with colder surfaces and subsequent dilution is done in a manner in which temperature is diminished gradually. Two different protocols are normally used for this purpose:

1. One step dilution with either preheated diluter (~33 °C) or diluter at room temperature.
2. A two steps dilution with first a 1:1 dilution with preheated diluter (~33 °C), followed by a second dilution in either a preheated diluter or a diluter kept at room temperature.

After the final dilution, filling of commercial doses is done and the semen is allowed to cool down gradually to 17 °C. When semen doses are to be transported, special precautions are taken to avoid temperature fluctuations. Further storage of diluted semen is done at 17 °C, at which temperature semen metabolism is reduced, a condition necessary to extend storage time.

Storage Medium

The storage media for liquid boar semen aim to prolong sperm survival, to provide energy to the cells, to buffer the pH of the suspension and to avoid the growth of bacteria. Therefore, porcine semen extenders contain ions to maintain the osmotic pressure of the medium, glucose as energy source, buffering systems to stabilize the pH of the extender and EDTA and antibiotics to prevent bacterial overgrowth. The presence of glucose as the only energy source and the low oxygen content in the recipient in which diluted semen is stored stimulates the glycolytic metabolism. Consequently, the intracellular pH of spermatozoa is lowered which reduces their motility and enables them to survive several days at ambient temperature. Glucose also contributes largely to the osmotic equilibrium. The ions in the media for liquid boar semen are merely sodium bicarbonate and sodium citrate and are simultaneously used as buffer. In BTS extender, also KCl is added to prevent the potassium loss from inside the cells, and subsequent loss of motility due to Na-K pump inefficacy. Porcine spermatozoa are rather tolerant to minor changes in osmolality of the extender. Iso-osmotic and slightly hyper-osmotic media are preferred for optimal preservation of fertilizing capacity. Incubation in media below 250 mOsm and above 300 mOsm rendered irreversible damage to the membranes and subsequent loss of motility.

EDTA is added for its chelating properties. When Ca-ions are captured, the initiation of capacitation is inhibited. As a consequence, the fertilizing capacity of the spermatozoa is preserved. Depending on the composition of the extender, semen can be stored for 2 to 3 days in short-term extenders and up to five days or longer in long-term extenders. Long-term extenders differ from short-term extenders mainly by the use of complex buffering systems (HEPES, Tris), mostly in addition to the bicarbonate buffering system, and by the presence of Bovine Serum Albumin (BSA). The latter has a positive influence on sperm survival due to the absorption of metabolic bacterial products from the extender. Cysteine is used as a membrane stabilizer inhibiting capacitation.

To prevent bacterial proliferation during storage, antibiotics are added to the extender. Bacteria originate mostly from the prepuce, thus depending on the semen collection technique, from semen manipulation or from the water used in the extender preparation. Depending on the species, bacteria have deleterious effects on semen quality, namely depressed motility, cell death and agglutination, either by direct effect on the spermatozoa or by acidifying the environment. European legislation prescribes an antibiotic combination equivalent to 500 IU/ml penicillin, 500 IU/ml streptomycin, 150 mg/ml lincomycin and 300 mg/ml spectinomycin, for having a broad antibacterial spectrum and activity towards leptospira. In practice most commercial extenders use aminoglycosides, especially gentamycin. However, bacterial contamination should be first minimized by good hygiene and general sanitation by personnel.

The extender-concentrates are diluted in distilled or de-ionized water. Next to the bacterial quality of the water, the electrolyte content, especially the absence of calcium ions, is an important characteristic for the water used to make the extender.

The comparison of different semen extenders has been subject of two kinds of studies: studies comparing different extenders in vitro, focusing on quality of semen after storage and studies comparing fertility in vivo after insemination of semen stored for several days or stored in different extenders.

In vitro experiments showed no differences in cell viability between short-term and long-term extenders during 9-day storage. Motility remained unchanged within the first 72 hours, even in BTS-extenders, the most widely used short-term extender. Based on in vivo results, differences were noticed between different extenders but it was not always possible to relate these differences to the type of extender. Differences between long-term extenders were observed from day 4 of storage onwards. The limited number of extenders compared in each study makes it difficult to set up a ranking of the semen preserving quality of long-term semen extenders.

Storage of Porcine Semen in Frozen State

As mentioned above, porcine spermatozoa are particularly sensible to low temperatures and to rapid cooling due to the specific composition of the cell membrane. Cold shock can be solved technically by inducing cold resistance, namely incubating sperm at ambient temperature for several hours, by contact with seminal plasma that has a protective effect on spermatozoa, together with controlled freezing protocols. The variation in freezability of individual boar's semen is however more difficult to solve.

Semen extenders for frozen boar semen are completely different from extenders for liquid semen. The presence of egg-yolk, containing low density lipoproteins and cholesterol, has a protective effect on sperm membrane during cooling. Cryoprotectants, especially glycerol are added in low concentration to the medium in order to diminish membrane damage by freezing. Additionally, sugars and synthetic detergents are added, the latter having a modifying effect on the egg yolk inducing a better membrane stability of the cell membrane. Thawing of the semen dose has been another point of concern. Thawing has to be fast in order to maintain sperm motility and acrosome integrity afterwards. Both processes i.e. freezing and thawing result in plasma membrane changes, explaining the variety of protocols available.

The fertility results with frozen semen have improved: cervical insemination results in a 75% farrowing rate and a litter size of 9.6. Nevertheless, freezing and thawing are time consuming processes, restricting the use of frozen semen for specific indications, such as long transport times and conservation of valuable genetic material.

Insemination Strategies

The management of AI is very important to determine the success of the procedure and the reproductive performance of the sows. Estrus control, timing and number of inseminations, the technique of AI, semen storage on farm and the use of new AI technologies, all require a specialized knowledge of pig reproductive physiology. The following measures could be taken to optimize the efficiency of AI on pig herds.

Boar Stimuli

Correct timing of insemination requires careful detection of oestrus at regular intervals. Boar stimuli are important in promoting follicular development and expression of oestrus behaviour. Additionally, a high level of boar stimuli increases the frequency of uterine contractions, indicating a supportive role for passive sperm transport through the long uterine horns at the time of insemination. This effect can only be partially mimicked by a robot teaser boar which emits olfactory, acoustic and visual boar cues. Increase of oxytocin concentrations in peripheral blood plasma occurs in immediate response to boar presence and lasts for approximately 10 min. Therefore, exposure of sows to a boar during both back pressure testing and insemination is crucial.

Timing of Insemination

Many studies have investigated time-relationships between oestrus, ovulation, insemination and fertilization using ultrasound testing. The key observation is that ovulation occurs at the beginning of the last third of oestrus regardless of the overall duration of oestrus. Precise prediction of the time of spontaneous ovulation in individual pigs has not yet been achieved. However, prediction of oestrus duration by observing the onset of oestrus after weaning has found broad acceptance in AI practice for calculation of the expected time of ovulation. AI should be timed as close as possible to ovulation, preferably within 12 to 24 h before ovulation. The benefit of ultrasound testing of ovarian morphology for pig fertility management has been shown in practice. Determination of the time of ovulation in relation to oestrus behaviour and AI management in representative numbers of sows on consecutive days has a great potential to provide short cuts in AI timing and to develop farm-specific strategies for improvement of AI management.

Use of New AI Technologies

The development of techniques to inseminate with low numbers of spermatozoa in a small volume has increased insemination efficiency. This is particularly interesting when using spermatozoa of high value that are impaired, e.g. by freezing and thawing or sex-sorting. Post-cervical or intrauterine insemination with several devices has been developed to traverse the cervix and deposit sperm in the uterine body or posterior horn of multiparous sows. Compared to standard transcervical AI, post-cervical AI allows a threefold reduction in the numbers of spermatozoa to be inseminated, whereas deep intrauterine AI allows a 5 to 20 fold reduction. The use of post-cervical insemination varies among and within countries. Limits may arise from the use in sows only, skills needed for catheter handling, and the possibility of damaging cervical or uterine tissue. Semen encapsulation in a barium alginate membrane has been demonstrated to allow a single insemination. Laparoscopy offers the possibility of inseminating a very low number of spermatozoa (i.e. 0.3×10^6) into the oviduct in anaesthetized pigs. However, the risk of polyspermic fertilization is substantial. Due to surgical intervention, its use is not appropriate in practice.

Embryo Transfer

Animal Embryo Breeding is an asexual reproduction of specific oocytes or embryos artificially by current developed biotechnology. The Science of Animal Embryo Breeding is to study how to use the embryo manipulation technologies to improve, create, and clone new animal individual or breed. Current developed techniques include nuclear transfer, cytoplasmic transfer or replacement, in vitro fertilization (IVF), sperm cytoplasmic injection (ICSI), parthenogenesis and androgenesis, embryo cloning, sex selection, transgenesis, gene knock out, stem cells and somatic cell cloning, etc. Although embryo breeding is a branch of traditional animal breeding discipline, the science of animal breeding is concerned with the application of the principles of population genetics and qualitative genetics to the improvement of domestic animals. However, Animal Embryo Breeding is concerned with application of the developed embryo biotechnologies to new animal individual creation, genetic cloning and preservation of animal breeds.

Superovulation (Multiple Ovulations)

In the normal reproductive cycle of a non-pregnant female, one ovarian follicle (out of the 20 that develop) matures and ruptures, releasing one fertile egg at a time. The time of ovulation varies in different animals—21 days for cow and horse; 16 days for sheep and goat.

The circulating gonadotrophic hormone is closely associated with ovulation and release of egg. By increasing the concentration of this hormone, more ovarian follicles can be induced to ripen and produce more eggs. This process, known as superovulation or multiple ovulations may yield not less than 8-10 eggs at a time. Some animal breeders were successful in super ovulating animals to yield as many as 60 eggs at a time. This largely depends on the breed, nutrition and health of the animal, besides the environmental factors.

By administering prostaglandin F_{2a} (PGF_{2a}) and follicle stimulating hormone (FSH), estrous can be induced. Artificial insemination is carried out in the super ovulated females. AI is preferred to natural mating, since the purpose of superovulation is to genetically improve the progeny. As the eggs are fertilized, they undergo development to form embryos.

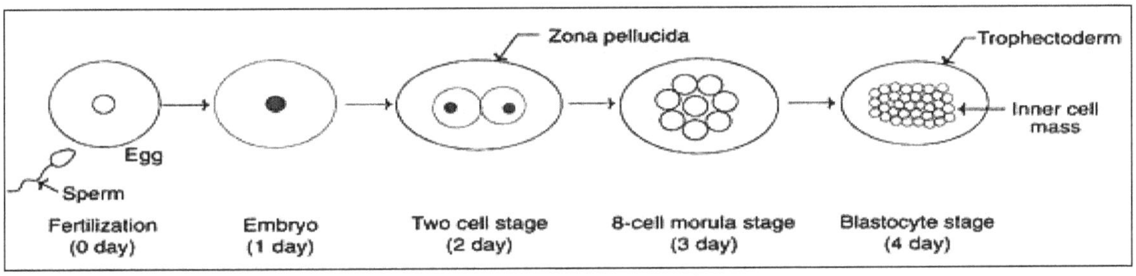

The early stages of embryo development.

Multiple Ovulation with Embryo Transfer (MOET)

Sometimes, the process of multiple ovulations (superovulation described above) and embryo transfer are considered together which is referred to as MOET. The embryos developed in the super ovulated animals (described above) are recovered after 6-8 days of insemination. In case of cattle, the recovery process is easy and can be done by using a catheter. For certain animals with smaller reproductive tract, surgical procedures may be needed to expose the oviduct and recover the embryos. These embryos are examined microscopically and identified.

The embryos are then transferred into a synchronized recipient (i.e. the foster mother) by using a procedure, which is comparable to artificial insemination. Alternately, the embryos can be frozen and stored for use at an appropriate time and place later. About 50-60% of pregnancy could be achieved in cattle with transferred embryos. Thus, a super ovulated female may result in 5-6 pregnancies. The embryos may also be subjected to manipulations, as described.

Embryo Splitting

It is possible to increase the number of progenies (almost to double) by splitting the embryos. In addition, embryo splitting results in identical twins for various purposes (particularly genetic research). Embryo splitting techniques have been refined, and are routinely used these days. The embryos are suspended in a hypertonic (high osmotic) sucrose and bovine serum albumin (BSA) containing culture medium. This results in shrinking of embryo cells and settling of the embryos to the bottom of the container (usually a petridish) due to increased density.

Further, the embryos stick to the bottom due to electrostatic interaction of the negative charges on it (due to attachment of albumin) and the positive charges of the petridish. The inner cell mass (ICM) of the blastocyte can be bisected by using a surgical blade and a micromanipulation technique (with the help of an inverted microscope). Splitting of each embryo results in two equal halves.

As the embryos are split, each hemispherical cell mass reforms spheres. These split embryos can be transferred into the oviducts of synchronized recipients. The pregnancy rate with split embryos is about 5-10% less than the theoretical calculations. Thus, about 9 progenies (instead of only 5 in the original embryo transfer) can be developed through MOET. By MOET, it is possible to produce at least four identical offspring's in animals at a time.

Embryo Biopsy

Embryo biopsy involves the removal of a few cells from the embryo (mostly from the trophoblastic cells of the trophectoderm) for analysis to determine the sex. In recent years, this technique is becoming popular for the detection of genetic diseases. By using embryo biopsy, it is possible to stop the transfer of embryos with genetic abnormalities and undesirable traits.

Embryo Sexing

Before the transfer or implantation of the embryo, the progeny sex can be determined. Embryo sexing is based on the principle of detecting the presence or absence of Y chromosome in the embryo biopsy cells. The absence of Y chromosome indicates that the embryo is a female.

In recent years, the presence DNA sequences specific to Y chromosome are being detected for embryo sexing. Another development is the use of polymerase chain reaction to amplify the DNA sequence (of Y chromosome) even from a single cell and determination the sex.

Limitations of Embryo Transfer

1. The supply of embryo from super ovulated donors is limited.
2. Freezing and thawing of embryos requires a lot of care to keep them functionally intact.
3. Embryo transfer requires technical skill, besides high cost factor.

Application of Embryo Breeding in Animal Improvement

Genomic Reconstruction by Somatic Cloning and Parthenogenesis

When a bull or cow with elite production performance in beef cattle population is discovered, the breeding aim will be to accelerate this cow or bull reproduction to propagate a new breed of cattle. By normal breeding mating, this cow may lose half its inherent genes in its offspring. However, by the means of cloning techniques, many individuals of the same genotype can be theoretically produced. Thus, the accuracy of evaluation may be greatly increased. In spite of low cloning efficiency, many scientists are still interested in animal cloning techniques, which will eventually be used to clone very valuable animals, such as breeding stock, transgenic animals, and endangered species.

By the means of cell nucleus transfer technology, a new animal can be produced using androgenesis method. Androgenesis is a male parthenogenesis in which only paternal chromosomes are kept in the embryo with the removal of the egg nucleus at the fertilization. This is a reproductive pattern from two male parents. After an oocyte nucleus has been removed, a male diploid cell is transferred into this egg in which the oocyte cytoplasm will induce this diploid cell going through meiosis to become a haploid MII oocyte. After inducement, a male sperm is injected into this oocyte to produce a paternal embryo. Finally, this modified embryo will be transferred into receipt cow to produce a new individual bull with two male parents.

Create New Genetic Variation in Population by Genomic Modification

The current animal breeding strategies are mainly based on the principle of selective breeding including the morphology of animal body, the application of quantitative genetics theory, the estimation of breeding value by phenotype, pedigree, BLUP (best linear unbiased prediction) method, and genetic markers. These methods mainly add genetic improvement by increasing the frequency of advantageous alleles of many loci, but actually very few of gene loci are identified. These techniques do not change gene movement from different species or genera due to reproductive barrier, while the new developed transgenic technique can remove the breeding barriers between different species or genera.

The most efficient method of transgenesis in mammals is the genetic manipulation of the pronuclear stage embryo. By injecting foreign DNA into one of the two pronuclei of the zygote, the birth offspring may contain a functional foreign gene in the genome. In the last 20 years, many kinds of transgenic species have been produced for agriculture and medicine application. For example, the transgenic technology in beef cattle industry may improve animals for faster growth, higher quality beef products, or disease resistance.

The transgenesis first starts with identification of the genes of interest. Current molecular biotechnology may help us to search for some interesting markers used as reference points for mapping relevant genes. These molecular markers can also be used for identification of the animals carrying the transgenes. Most of the quantitative genetic loci (QTL) are polygenic in nature but the manipulation of transgenesis is a single gene trait. The technology holds promises in the future in moving polygenic QTL across the breeding barriers of animals. However, it is expected that molecular markers will serve as a potential tool to geneticists and breeders to evaluate the existing germplasm, and to manipulate it to create animals of desired traits.

Shorten Generation Interval by Embryo in Vitro Production

As the oocyte in vitro maturation (IVM) and in vitro fertilization (IVF) techniques rapidly develop, the ultrasound-guided oocyte retrieval (TVOR) or nonsurgical ovum pick up (OPU) technique can retrieve many oocytes repeatedly from a cow or a heifer. As many as 1000 oocytes have been collected from one female cattle in a year. Thus, the embryo in vitro production (IVP) technology has been able to promote a cow to produce more than one hundred offspring in a year and greatly accelerate herd genetic improvement speed. In order to improve ordinary cattle herd, slaughterhouse ovaries also may be used as in vitro embryo production. A lot of oocytes could be obtained from slaughter house cow ovaries. After maturation, these oocytes may be inseminated with elite bull semen for in vitro fertilization. Although the detail genetic backgrounds of these slaughterhouse animals are not known, these embryos have a very high genetic merit from elite bulls. Using these embryos, an ordinary cow herd could obtain at least 50% genetic improvement.

The multiple ovulation and embryo transfer (MOET) was used initially to produce more embryos from genetic elite cows in shorter time periods. Currently, the MOET breeding schemes have widely established in many countries and their use accounts for about 80% of cattle embryos transferred commercially. Currently, the application of transvaginal ultrasonically guided OPU technique may significantly improve MOET scheme efficiency because about 1000 oocytes may be collected and 300 embryos may be produced in vitro from a cow in a year at frequent intervals using IVF technology. Also, oocytes may be collected from prepubertal heifers and cattle generation interval may be shorted for 2-3 years. The combination of MOET program with OPU/IVF technique is providing a more efficient way to produce more embryos from an individual donor donor than superovulation stimulation program. Thus, OPU/IVF technique greatly increases MOET breeding scheme efficiency in milk and beef industry.

Increased Economy from Animal Population by Sex Selection

Animal sex selection may increase animal economical value for humans. Embryo breeding theory may provide several ways for animal sex selection, including sperm sex selection and preimplantation embryo sex selection. Sperm sex selection is to try to separate semen into X- or Y-bearing chromosome sperm by flow cytometry. Current sorted sperm has been successfully used in IVF for in vitro embryo production and artificial insemination in cattle.

Another sexing pathway is to determine the sex of an embryo prior to transfer. Preimplantation genetic diagnosis (PGD) technique has become an efficient method for sex selection. Y-specific chromosome probe for polymerase chain reaction (PCR) and Fluorescent In Situ Hybridization (FISH) are two common methods in animal sex determination. On the ordinary farm, cattle embryos may be sexed by complete cell biopsy and PCR technique. Our clinic farm practice showed that a few of trophoectoderm cells could be microbiopsied from blastocyst embryos by transzonal incision using a microsurgical blade. The mini-tube PCR was carried out for 30 minutes and the gel electrophoresis was run for 20 minutes. The sexing result could be obtained in 2 hours. These results clearly demonstrate that the microsurgical technique and subsequent PCR sex analysis allow the rapid commercial exchange of genetic resources on the basis of fresh or frozen sex-desired embryos in embryo transfer programs.

Fluorescent in situ hybridization (FISH) technique has also been used as embryo chromosome set (karyotype) diagnosis. A blastomere is removed from an embryo by micromanipulation, and then used to examine the embryo X/Y chromosomes by FISH. Recently, new developed technologies in PGD allow examining of all chromosomes and identifying certain genes or genetic mutations, such as the competitive genomic hybridization (CGH) and microarray analysis. More recently, novel developed Next Generation Sequencing (NGS) for preimplantation genetic screen (PGS) is now being offered clinically to provide comprehensive, accurate screening of all 24 chromosomes for selections of euploid embryos. PGS results generated are comparable to those achieved with the CGH technology, with improved accuracy, sensitivity, and resolution for more accurate detection of euploid embryos, aneuploidies, chromosome imbalances (translocations), and embryo mosaicism. NGS is a superior technology because it looks at close to 1.1 million data points on the genome compared to around 3,000 with CGH.

Preservation Breeding

Many animal breeders are interested in preserving bloodlines of animals, either of a rare breed, or of rare pedigrees within a breed. Therefore, Rahbek put forward a preservation breeding concept to describe the purpose of preservation breeding, which is to protect genetic diversity within a species, and to preserve valuable genetic traits that may not be popular or in fashion in the present, but may be of great value in the future. In the animal embryo breeding program, two kinds of cells including reproductive cells and somatic cells may be cryopreserved in liquid nitrogen for future use. Reproductive cell cryopreservation is an important branch of embryo breeding science because it involves the preservation of gametes (sperm and oocytes), embryos, and reproductive tissues (ovarian and testicular tissues) for future use in the assisted reproductive technology. Practically, animal embryo breeding program may provide a sperm and embryo bank with the objective of avoiding genetic dilution and irreplaceable gene losses of the valuable "naturalized breeds" germplasm. It is much lower in cost than normal animal breeding, preserving rare native animal breed plan. At present, many countries have set up gene banks to store frozen embryos and semen of various animal species including native cattle, pig, and some endangered animals.

The development of embryo freezing technologies has revolutionized cattle breeding. Since then, advancements in cryobiology, cell biology, and domestic animal embryology have enabled the development of embryo preservation methodologies for our other domestic animal species, including sheep and goats. Currently, use of preserved embryos has become a routine breeding alternative for all domestic animal species. This freezing and storage methodology may provide for maternal germplasm, global genetic transport, increased selection pressure of herd genetics, and genetic resource rescue.

In the conventional breeding program, an outstanding bull may maintain normal mating for 5 years. However, if this bull semen is cryopreserved, it will extend the bull's breeding time. In embryo breeding program, when some elite bulls leave very few sperm, we may use intra-cytoplasmic sperm injection (ICSI) technique to inject a single sperm to an oocyte so that genetic merit embryos are obtained. Also, sperm cell genome cloning technique may be used to produce many copies of a specific sperm. The application of this technique to beef and dairy cattle industry has greatly increased merit bull spread in animal herd.

Like normal reproduction, somatic cell nuclear transfer (SCNT) starts with an egg or oocyte, but here the nucleus of the egg needs to be removed. Then the nucleus from a somatic (skin) cell is transferred into the enucleated egg which would be analogous to the sperm entering the oocyte. As this develops into a blastocyst, cells from the inner cell mass can be isolated and purified to serve as a source for pluripotent stem cells. In animal embryo breeding, somatic cell is also an important genetic resource. Therefore, the somatic cells, such as skin, hair, and other cells from rare and endangered animals may be collected and cryopreserved so that they can be used in the future.

In Vitro Fertilization

The limitations of embryo transfer could be successfully overcome by using in vitro fertilization (IVF). IVF basically involves fertilization of oocytes of a female animal in the laboratory conditions under artificial conditions. This is in contrast to the natural fertilization, which occurs in the uterus. IVF is reasonably successful as it results in 70-80% of the fertilized eggs.

Through IVF technology, a large number of offspring's can be produced from a single animal.

Thus, a female animal, which normally produces 4-5 offspring's in her lifetime, can produce as many as 50-20 offspring's by employing IVF. In vitro fertilization involves the following stages:

Oocyte Recovery

During the course of estrous cycle, the ovarian follicles grow, get filled with fluid and become Graafian (antral) follicles. The oocytes from these follicls can be recovered by using laparoscopic surgery. It is possible to recover more number of oocytes from super ovulated donors.

In Vitro Maturation (IVM) of Oocytes

The immature oocytes recovered in the above step, when incubated in vitro will result in mature oocytes (eggs). The occurrence of a large number of antral follicles throughout reproductive life in the ovaries of cattle and sheep are known.

It is theoretically possible to remove them by biopsy and induce oocyte maturation in vitro. This may ultimately result in the production of hundreds and thousands of eggs from a single female. However, the potential of in vitro maturation of oocytes is limited for various reasons-degeneration of follicles, unknown metabolic and hormonal conditions.

Fertilization of Eggs

In vitro fertilization of the eggs is carried out by using semen obtained from a superior male animal. For IVF, the eggs are carried in small droplets (micro droplets) of culture medium. Each micro droplet usually carries about 10 eggs. A dose of sperm approximately one million cells per ml is adequate for IVF. The penetration of the sperm into the egg is facilitated by supplementing the medium with certain compounds (penicillamine, epinephrine etc.).

Embryo Culture

The IVF embryos must be maintained in the in vitro conditions for a few days (about 7 days for sheep and goat, 8 days for cattle). This allows the development of embryos to blastocyte stage. Approximately, 60% of the IVF embryos in vitro culture can form blastocytes.

Implantation of Embryos

The 7 or 8 day old embryos from the in vitro culture are implanted in the reproductive tract of the recipient female which acts as a foster mother or surrogate mother.

Limitations of IVF

The pregnancy loss of IVF embryos, particularly during the first two months, is very high, although the reasons are not clearly known. The following parameters, which may be considered as limitations of IVF, may contribute to fetal losses of IVF embryos:

1. Genetic defects in oocytes.
2. Genetic defect in fertilizing sperms.
3. Environmental mutagenesis.

4. Inadequate supply of nutrients and hormones.
5. Exposure to toxic agents and free radicals.

In recent years, some improvements have been made in IVF to minimize the fetal losses.

Embryo Cloning

A clone is an organism that is descended from, and genetically identical to, a single common ancestor. Animals can be cloned by embryo splitting or nuclear transfer. Embryo splitting involves bisecting the multicellular embryo at an early stage of development to generate "twins". This type of cloning occurs naturally and has also been performed in the laboratory with a number of animal species.

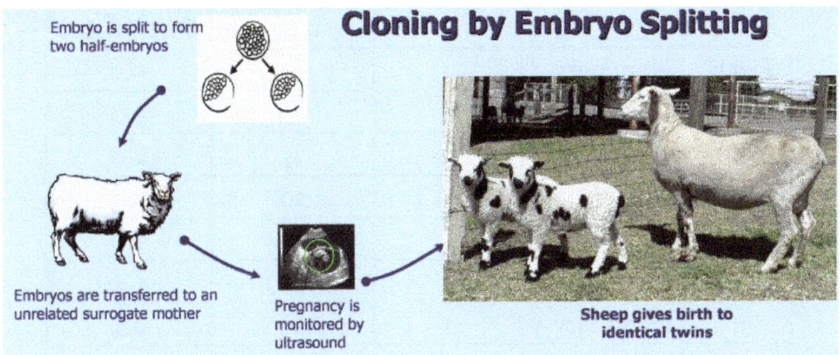

Cloning can also be achieved by nuclear transfer where the genetic material from one cell is placed into a "recipient" unfertilized egg that has had its genetic material removed by a process called enucleation. The first mammals were cloned via nuclear transfer during the early 1980s, almost 30 years after the initial successful experiments with frogs. Numerous mammalian species have been cloned from cells of pre-implantation embryos: namely mice, rats, rabbits, pigs, goats, sheep, cattle and even two rhesus monkeys.

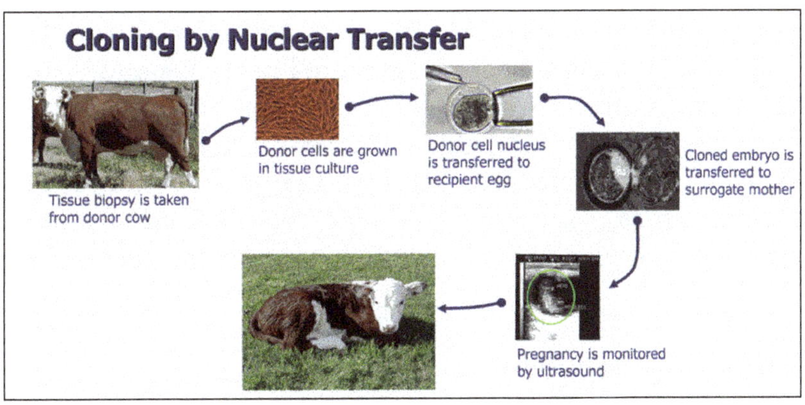

Dolly the sheep was the first animal to be cloned via nuclear transfer from a cultured adult cell in 1996. Since then, a diverse range of adult tissues have been successfully cloned in a variety of species including cattle, mice, pigs, cats, rabbits, goats, and zebrafish.

The proportion of adult cell nuclei to develop into live offspring after transfer into an enucleated egg is very low. High rates of abortion have been observed at various stages of pregnancy after placement of the eggs containing the adult cell nuclei into recipient animals. Various abnormalities have been observed in cloned cows and mice after birth and this has been found to be somewhat dependent on the type of tissue that originated the nuclei used to make the clone. The reasons for the low efficiency of cloning by nuclear transfer are currently under investigation but it is thought that it may be related to insufficient nuclear reprogramming as the cloned nuclei goes from directing the production of an adult somatic cell to directing the production of a whole new embryo.

Mammals Cloned From Adult Cells

Donor		Newborns		
Species	Cell/Tissue	No.	% Embryos Transferred	% Oocytes Constructed
Sheep	Mammary	1	3	0.2
Cattle	Cumulus	5	83	5.0
	Oviduct	3	75	2.0
	Granulosa	10	10	1.0
	Muscle	4	15	2.0
	Leukocyte	1	2	0.1
	Mammary	1	25	0.5
	Ear	1	6	0.1
	Skin	1	4	0.5
	Skin	6	11	1.8
Mouse	Cumulus	23	2	1.0
	Tail	3	1	0.4
	Sertoli	16	4	4.0
Pig	Granulosa	5	7	3.0

Cloning offers the opportunity to make transgenic animals from cultured cells that have been genetically engineered. The first genetically engineered or transgenic mammalian clones were sheep born in 1997 carrying the coding sequences for human clotting factor IX, which is an important therapeutic for hemophiliacs. One of these transgenic sheep, Dolly, expressed this protein in her milk. Cloning may also be useful for the preservation of rare and endangered species. In human therapeutics, patients may be able to clone their own nuclei to make healthy tissue that could be used to replace diseased tissue without the risk of immunological rejection.

Making Genetically Engineered Clones

Fetal cells in culture were transfected with a DNA sequence containing a selectable marker (neomycin resistance), the human gene for clotting factor IX, and a regulatory sequence to direct the gene to function only in the mammary gland. Following selection for neomycin resistance, nucleus from surviving cells were each transferred to an enucleated egg. Of the three transgenic clones born, one named Dolly survived and later secreted human clotting factor in her milk. Dolly is the first transgenic mammalian clone.

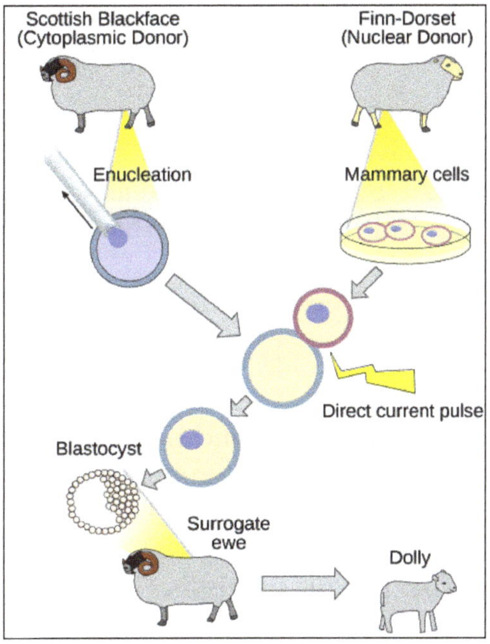

Transgenic Animals

The dependence of man on animals such as cattle, sheep, poultry, pig and fish for various purposes (milk, meat, eggs, wool etc.) is well known. Improvement in the genetic characteristics of livestock and other domestic animals (e.g., high milk yield, weight gain, etc.), in the early days, was carried out by selective breeding methods.

This technique primarily involves a combination of mating and selection of animals with improved genetic traits. Although selective breeding is very time consuming and costly, it was the only method available, till some years ago, to enhance the genetic characteristics of animals.

For larger animals with long gestation period, it might take several decades to create a desired character by conventional breeding. With the advent of modern biotechnology, it is now possible to carry out manipulations at the genetic level to get the desired characteristics in animals.

Trans-genesis refers to the phenomenon of introduction of exogeneous DNA into the genome to create and maintain a stable heritable character. The foreign DNA that is introduced is called transgene. And the animal whose genome is altered by adding one or more transgenes is said to be transgenic.

The transgenes behave like other genes present in the animals' genome and are passed on to the offspring's. Thus, transgenic animals are genetically engineered or genetically modified organisms (GMOs) with a new heritable character. It was in 1980s, the genetic manipulation of animals by introducing genes into fertilized eggs became a reality.

Importance of Transgenic Animals-general

Trans-genesis has now become a powerful tool for studying the gene expression and developmental processes in higher organisms, besides the improvement in their genetic characteristics. Transgenic animals serve as good models for understanding the human diseases.

Further, several proteins produced by transgenic animals are important for medical and pharmaceutical applications. Thus, the transgenic farm animals are a part of the lucrative world-wide biotechnology industry, with great benefits to mankind. Trans-genesis is important for improving the quality and quantity of milk, meat, eggs and wool production, besides creating drug resistant animals.

Milk as the Medium of Protein Production

Milk is the secretion of mammary glands that can be collected frequently without causing any harm to the animal. Thus, milk from the transgenic animals can serve as a good and authenticated source of human proteins for a wide range of applications. Another advantage with milk is that it contains only a few proteins (casein, lactalbumin, immunoglobulin etc.) in the native state, therefore isolation and purification of a new protein from milk is easy.

Commonly used Animals for Trans-genesis

The first animals used for trans-genesis was a mouse. The 'Super Mouse', was created by inserting a rat gene for growth hormone into the mouse genome. The offspring was much larger than the parents. Super Mouse attracted a lot of public attention, since it was a product of genetic manipulation rather than the normal route of sexual reproduction. Mouse continues to be an animal of choice for most transgenic experiments. The other animals used for trans-genesis include rat, rabbit, pig, cow, goat, sheep and fish.

Position Effects

Position effect is the phenomenon of different levels of gene expression that is observed after insertion of a new gene at different position in the eukaryotic genome. This is commonly observed in transgenic animals as well as plants. These transgenic organisms show variable levels and patterns of transgene expression. In a majority of cases, position effects are dependent on the site of transgene integration. In general, the defective expression is due to the insertion of transgene into a region of highly packed chromatin. The transgene will be more active if inserted into an area of open chromatin.

The positional effects are overcome by a group of DNA sequences called insulators. The sequences referred to as specialized chromatin structure (SCS) are known to perform the functions of insulators. It has been demonstrated that the expression of the gene is appropriate if the transgene is flanked by insulators.

Animal Bioreactors

Trans-genesis is wonderfully utilized for production proteins of pharmaceutical and medical use. In fact, any protein synthesized in the human body can be made in the transgenic animals, provided that the genes are correctly programmed. The advantage with transgenic animals is to produce scarce human proteins in huge quantities. Thus, the animals serving as factories for production of biologically important products are referred to as animal bioreactors or sometimes pharm animals. Frankly speaking, transgenic animals as bioreactors can be commercially exploited for the benefit of mankind.

Once developed, animal bioreactors are cost- effective for the production of large quantities of human proteins. Routine breeding and healthful living conditions are enough to maintain transgenic animals.

Transgenic Animals in Xenotransplantation

Organ transplantation (kidney, liver, heart etc.) in humans has now become one of the advanced surgical practices to replace the defective, non-functional or severally damaged organs. The major limitation of transplantation is the shortage of organ donors. This often results in long waiting times and many unnecessary deaths of organ failure patients.

Xenotransplantation refers to the replacement of failed human organs by the functional animal organs. The major limitation of xenotransplantation is the phenomenon of hyper acute organ rejection due to host immune system.

The organ rejections is mainly due to the following two causes:

1. The antibodies raised against the foreign organ.
2. Activation of host's complement system.

Pigs in Xenotransplantation

Some workers are actively conducting research to utilize organs of pigs in xenotransplantation. It is now identified that the major reason for rejection of pig organs by primates is due to the presence of a special group of disaccharides (Gal-α 1, 3-Gal) in pigs, and not in primates.

The enzyme responsible for the synthesis of specific disaccharides in pigs has been identified. It is α 1, 3-galactosyltransferase, present in pigs and not in primates. Scientists are optimistic that knockout pigs lacking the gene encoding the enzyme α 1, 3-galactosyltransferase can be developed in the next few years. Another approach is to introduce genes in primates that can degrade or modify Gal-α 1, 3-Gal disaccharide groups (of pigs). This will reduce immunogenicity.

Besides the above, there are other strategies to avoid hyperactive organ rejection by the hosts in xenotransplantation.

1. Expression of antibodies against the pig disaccharides.
2. Expression of complement— inactivating protein on the cell surfaces.

By the above approaches, it may be possible to overcome immediate hyperactive rejection of organs. The next problem is the delayed rejection which involves the macrophages and natural killer cells of the host.

Another concern of xenotransplantation is that the endogenous pig retroviruses could get activated after organ transplantation. This may lead to new genetic changes with unknown consequences.

The use of transgenic animals in xenotransplantation is only at the laboratory experimental stages, involving animals. It is doubtful whether this will become a reality in the near future. There is a vigorous debate concerning the ethics of xenotransplantation and the majority of general public are against it.

References

- Animal-breeding, science: britannica.com, Retrieved 20 February, 2019
- Artificial-insemination, breeding-and-genetics, animal-science: animalsmart.org, Retrieved 27 June, 2019
- Particularities-of-bovine-artificial-insemination, artificial-insemination-in-farm-animals: intechopen.com, Retrieved 7 January, 2019
- Artificial-insemination-in-pigs, artificial-insemination-in-farm-animals: intechopen.com, Retrieved 3 April, 2019
- Embryology-animal-embryo-breeding, embryology: intechopen.com, Retrieved 13 July, 2019
- Artificial-breeding-of-animals-4-approaches, reproductive-technology: biologydiscussion.com, Retrieved 11 February, 2019
- Cloning: animalbiotech.ucdavis.edu, Retrieved 18 August, 2019
- Transgenic-animals, animals: biologydiscussion.com, Retrieved 25 May, 2019

Chapter 4

Animal Behavior

The different actions of animals such as their movement along with the underlying mental processes are studied within animal behavior. Some of the areas of interest within this field are animal learning, animal emotion and animal mimicry. This chapter discusses in detail these fields of study related to animal behavior.

Animal behaviour refers to the activities that constitute an animal's response to its external environment. Certain categories of behaviour are seen in all animals (e.g. feeding, reproduction) but these activities involve different movements in different species and develop in different ways. Some movements are highly characteristic of a species, whereas others are more variable and depend on the interaction between innate tendencies and learning during the individual's lifetime.

The study of animal behavior, called ethology. The study of animal behavior touches upon the fact that people receive joy from nature and also typically see themselves in a special role as stewards of creation. Behavior is one aspect of the vast diversity of nature that enhances human enjoyment. People are fascinated with the many behaviors of animals, whether the communication "dance" of honeybees, or the hunting behavior of the big cats, or the altruistic behavior of a dolphin. In addition, humans generally see themselves with the responsibility to love and care for nature.

The study of animal behavior also helps people to understand more about themselves. From an evolutionary point of view, organisms of diverse lineages are related through the process of descent with modification. From a religious point of view, human also stand as "microcosms of nature". Thus, the understanding of animals helps to better understand ourselves.

Ethologists engage in hypothesis-driven experimental investigation, often in the field. This combination of lab work with field study reflects an important conceptual underpinning of the discipline: Behavior is assumed to be adaptive; i.e., something that makes it better suited in its environment and consequently improves its chances of survival and reproductive success.

Ethology emerged as a discrete discipline in the 1920s, through the efforts of Konrad Lorenz, Karl von Frisch, and Niko Tinbergen, who were jointly awarded the 1973 Nobel Prize in Physiology or Medicine for their contributions to the study of behavior. They were in turn influenced by the foundational work of, among others, ornithologists Oskar Heinroth and Julian Huxley and the American myrmecologist (study of ants) William Morton Wheeler, who popularized the term ethology in a seminal 1902 paper.

One of the key ideas of classical ethology is the concept of fixed action patterns (FAPs). FAPs are stereotyped behaviors that occur in a predictable, inflexible sequence in response to an identifiable stimulus from the environment.

Kelp Gull chicks peck at a red spot on their mother's beak to stimulate the regurgitating reflex, another example of a fixed action pattern.

For example, at the sight of a displaced egg near the nest, the greylag goose (Anser anser) will roll the egg back to the others with its beak. If the egg is removed, the animal continues to engage in egg-rolling behavior, pulling its head back as if an imaginary egg is still being maneuvered by the underside of its beak. It will also attempt to move other egg-shaped objects, such as a golf ball, doorknob, or even an egg too large to have been laid by the goose itself.

Another important concept is filial imprinting, a form of learning that occurs in young animals, usually during a critical, formative period of their lives. During imprinting, a young animal learns to direct some of its social responses to a parent or sibling.

Despite its valuable contributions to the study of animal behavior, classical ethology also spawned problematic general theories that viewed even complex behaviors as genetically hardwired (i.e., innate or instinctive). Models of behavior have since been revised to account for more flexible decision-making processes.

Key Principles and Concepts

Behaviors are Adaptive Responses to Natural Selection

Because ethology is understood as a branch of biology, ethologists have been particularly concerned with the evolution of behavior and the understanding of behavior in terms of the theory of natural selection. In one sense, the first modern ethologist was Charles Darwin, whose book The Expression of the Emotions in Man and Animals has influenced many ethologists. (Darwin's protégé George Romanes became one of the founders of comparative psychology, positing a similarity of cognitive processes and mechanisms between animals and humans).

Note, however, that this concept is necessarily speculative. Behaviors are not found as fossils and cannot be traced through the geological strata. And concrete evidence for the theory of modification by natural selection is limited to microevolution—that is, evolution at or below the level of species. The evidence that natural selection directs changes on the macro-evolutionary level necessarily involves extrapolation from these evidences on the micro-evolutionary level. Thus, although scientists frequently allude to a particular behavior having evolved by natural selection in response to a particular environment, this involves speculation as opposed to concrete evidence.

Animals use Fixed Action Patterns in Communication

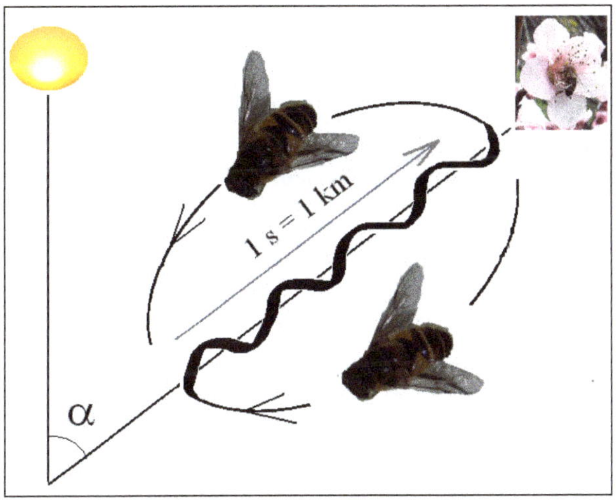

The honeybee's figure-eight dance is a fixed-action pattern that communicates information to other members of the group: the angle from the sun indicates the direction of a food source; the duration signifies its distance.

A fixed action pattern (FAP) is an instinctive behavioral sequence produced by a neural network known as the innate releasing mechanism in response to an external sensory stimulus called the sign stimulus or releaser. Once identified by ethologists, FAPs can be compared across species, allowing them to contrast similarities and differences in behavior with similarities and differences in form (morphology).

An example of how FAPs work in animal communication is the classic investigation by Austrian ethologist Karl von Frisch of the so-called "dance language" underlying bee communication. The dance is a mechanism for successful foragers to recruit members of the colony to new sources of nectar or pollen.

Imprinting is a Type of Learning Behavior

Imprinting describes any kind of phase-sensitive learning (i.e., learning that occurs at a particular age or life stage) during which an animal learns the characteristics of some stimulus, which is therefore said to be "imprinted" onto the subject.

The best known form of imprinting is filial imprinting, in which a young animal learns the characteristics of its parent. Lorenz observed that the young of waterfowl such as geese spontaneously followed their mothers from almost the first day after they were hatched. Lorenz demonstrated how incubator-hatched geese would imprint on the first suitable moving stimulus they saw within what he called a critical period of about 36 hours shortly after hatching. Most famously, the goslings would imprint on Lorenz himself (more specifically, on his wading boots).

Sexual imprinting, which occurs at a later stage of development, is the process by which a young animal learns the characteristics of a desirable mate. For example, male zebra finches appear to prefer mates with the appearance of the female bird that rears them, rather than mates of their own type. Reverse sexual imprinting has also observed: when two individuals live in close domestic proximity during their early years, both are desensitized to later sexual attraction. This phenomenon, known as the Westermarck effect, has probably evolved to suppress inbreeding.

Relation to Comparative Psychology

In order to summarize the defining features of ethology, it might be helpful to compare classical ethology to early work in comparative psychology, an alternative approach to the study of animal behavior that also emerged in the early 20th century. The rivalry between these two fields stemmed in part from disciplinary politics: ethology, which had developed in Europe, failed to gain a strong foothold in North America, where comparative psychology was dominant.

Broadly speaking, comparative psychology studies general processes, while ethology focuses on adaptive specialization. The two approaches are complementary rather than competitive, but they do lead to different perspectives and sometimes to conflicts of opinion about matters of substance:

- Comparative psychology construes its study as a branch of psychology rather than as an outgrowth of biology. Thus, where comparative psychology sees the study of animal behavior in the context of what is known about human psychology, ethology situates animal behavior in the context of what is known about animal anatomy, physiology, neurobiology, and phylogenetic history.
- Comparative psychologists are interested more in similarities than differences in behavior; they are seeking general laws of behavior, especially relating to development, which can then be applied to all animal species, including humans. Hence, early comparative psychologists concentrated on gaining extensive knowledge of the behavior of a few species, while ethologists were more interested in gaining knowledge of behavior in a wide range of species in order to be able to make principled comparisons across taxonomic groups.
- Comparative psychologists focused primarily on lab experiments involving a handful of species, mainly rats and pigeons, whereas ethologists concentrated on behavior in natural situations.

Animal Communication

Animal communication refers to the process by which one animal provides information that other animals can incorporate into their decision making. The vehicle for the provision of this information is called a signal. The signal may be a sound, colour pattern, posture, movement, electrical discharge, touch, release of an odorant, or some combination of these mediums.

Animals face daily decisions about how to behave. Choices can be as simple as a sea anemone deciding when to expand its tentacles or as complex as a male lion deciding whether to approach a reluctant mate. The decision, which may be reflexive or conscious, is guided by evolutionary biases based on alternative outcomes of choice, recent experience about likely conditions, and sensory information. An animal with access to complete information can always choose correctly. However, life is rarely so accommodating, and inputs often fail to provide complete information. Thus, communication is an important source of additional information that is incorporated into the decision-making process.

Signals are actions or anatomical structures whose primary function is the provision of information to another animal. However, not all actions by one animal that provide information to another animal qualify as signals. The noise created by a foraging mouse and used by an owl to locate and kill the mouse is a case in point. Mice have to feed, and the noises they create while feeding (e.g., through movement and chewing) are an inadvertent result of that activity. Thus, these sounds are not a signal. In contrast, the song of a wren is not inadvertent—wrens sing solely to communicate with other birds.

Cactus wrens (Campylorhynchus brunneicapillus) sing to communicate with other members of their species.

Senders and Receivers

An animal that provides a signal is called a sender. The animal to which the signal is directed is the receiver. The receiver uses the signal information to help make a decision. For example, if a receiver must choose either to fight with or to flee from an opponent, it brings to this decision biases and thresholds passed on to it by successful prior generations. This information helps the receiver avoid harm and find food, shelter, and mates. Prior experience in the receiver's own life may also play a role in shaping its evaluation of the situation. If it has routinely lost fights to larger animals, a useful strategy would be to assess the size of the opponent. This may be done by using vision or other means. For example, in some cases an opponent broadcasts a low-frequency sound signal at the receiver. Because only large animals can produce low-frequency sounds, this signal provides evidence that the opponent is large. The receiver integrates its perception of the sound frequency with its prior experience and inherited avoidance of harmful situations and thus decides to flee.

In this example, the receiver can interpret the signal only if it understands that low-frequency sounds tend to be associated with large body sizes. The association between alternative signals (e.g., sounds of different frequencies) and different alternative circumstances (e.g., relative sizes of opponents) is called a code. Codes can be characterized as probabilities that a sender will emit a given signal in any given circumstance. In a perfect code, only one signal will be used in a given context, and only one context will evoke that signal. Real codes do not need to be perfect, but they do need to be good enough that a receiver attending to signals makes better decisions than if it ignored the signals and relied only on other sources of information.

Animals differ widely in the mechanisms by which they acquire signal codes. Some codes are inherited genetically. For example, the sound-producing structures of many male insects generate a limited range of sound frequencies, and the ears of females are pretuned to be most sensitive to those frequencies. In other species, senders' sounds or body odours are determined by random genetic processes, and receivers must learn which signals go with which individuals. Many songbirds have genetic limits on the range of sounds they can sing, but they can learn one or more local variants within those limits during a short period in their youth. In certain species, such as parrots or humans, both sender and receiver must learn the appropriate vocal coding, and they can continue to learn alternative coding systems throughout life.

Different contexts require different kinds of information and thus different signals. The number of signals in a species' repertoire can range from 5 or 6 in the simplest non-social animals to 10–20 in social insects, such as bees and ants, or to 30–40 in social vertebrates, such as wolves and primates. Most animals produce signals to attract mates and then produce additional signals to synchronize mating. Signals for mediating conflicts, including signals of aggressive intention and signals of submission, are also widespread. In addition, territorial species require signals for declaring territory ownership, and in situations in which adults guard or feed their young, both parents and offspring require signals to coordinate parental care. Social animals may use signals to coordinate group movements, to assemble dispersed group members, or to display social affiliations. Some animals have special signals that they use to share food finds, to alert others about predator attacks, and even to alert approaching predators that they have been detected. In addition, bats, oilbirds, porpoises, and electric fish use the differences between their own emitted and subsequently received signals to extract information about the ambient environment. In many of these contexts, the relevant animal signals are designed to provide a receiver with ancillary information about the identity, sex, social affiliation, and location of the sender.

Wolves are social animals, and they thus require a large repertoire of signals to communicate different kinds of information.

Signal Production

The challenge faced by a sender is the creation of a controlled perturbation of the environment that can be detected and recognized by a receiver. Sound production is one mechanism. Sound travels in waves, and thus any sound can be characterized by its component frequencies and the physical size of each wave component (called the wavelength). The wavelength of a sound depends upon its frequency and the speed of sound in the propagating medium. The speed of sound is greatest in solids, intermediate in water, and least in air. Thus, a given frequency of sound in water has a wavelength 4.5 times longer than the same frequency in air, and the same frequency in a solid can be up to 15 times longer than that in air. This is important to animals using sound communication because it is physically difficult for an animal to produce a loud sound with a wavelength much larger than itself. For this reason, small animals tend to communicate with high-frequency sounds, and only large animals use low-frequency sound signals. Aquatic animals require higher-frequency signals than do similarly sized terrestrial animals.

Blue arrow-poison frogs (Dendrobates azureus) can communicate through sound production. Their bright colour also serves as a warning signal to predators.

The lowest frequencies that small insects, frogs, and birds can produce as signals may be many thousands of waves per second. Animal muscles cannot twitch this quickly, which makes sound production challenging. One solution is to use frequency multiplication. For example, hard-bodied animals drag a comb-like structure over a sharp edge. A single muscle contraction causes the sharp edge to hit successive teeth in the comb, thereby producing a sequence of sound waves. This is called stridulation. Arthropods all have hard exoskeletons, and by mounting the comb on one external body part and the sharp edge on the other, they can stridulate by rubbing the two hard parts together. For example, lobsters rub an antenna against the head, beetles rub a leg against the body, and crickets and katydids rub one wing over another. There are other techniques for frequency multiplication. Terrestrial vertebrates use muscles to force air into and out of their lungs while breathing. If thin membranes are inserted into this airflow, the membranes will flutter, producing sound waves at much higher frequencies than the airflow cycle. This is how frogs croak, lions roar, and birds sing.

In figure, Katydids use several different forms of communication. One of these forms is called stridulation and is characterized by the rubbing together of the insect's wings to create sound waves. These sound waves convey specific types of information and are detected by members of the same species.

To be able to use different sound signals in different contexts, animals must have some way to control and vary the sounds that they produce. Varying the rhythm of insect stridulation or bird breathing is one way to produce different signals. Air-breathing vertebrates can also change the tension on the vibrating membranes to produce quite complicated frequency modulations. A third mechanism is to produce sounds that initially contain many different frequencies and then selectively filter out some frequencies and amplify others. Animals such as katydids, frogs, bats, and howler monkeys have special resonating structures attached to their sound-producing organs that select the radiated frequencies and couple the sounds to the medium.

Bats have special resonating structures attached to their sound-producing organs that select specific sound frequencies. This enables them to use different sound signals in different contexts.

Light is another modality used for producing signals. Most visual signals rely on the presence of ambient light that is generated by the Sun. Similar to sound, light propagates as waves. When white light—which contains many different light frequencies—strikes an object, some of it is reflected, and it is this reflected light that creates a visual image of the object. If the object absorbs some frequencies of white light and reflects others, the receiver will see the object as coloured. When red light frequencies are absorbed, the object appears green or blue. When the green light frequencies are absorbed, the colour appears purple. Different animal groups tend to have different ranges of light frequencies that they can see. Birds, lizards, and some insects can see light frequencies well into the short-waved ultraviolet. Humans cannot see these short waves. Thus, in humans, objects that appear to be similar or uncoloured may be seen as different or highly coloured by a bird or a honeybee.

The challenge for a sender is to produce a visible image that is detectable against the background by a receiver. One way to do this is to move the signal body part in front of a static background or to move it in a different direction relative to a moving background. This simply requires normal muscle movements—no additional structure is needed. However, many senders enlarge or decorate the moved body part to increase the chances that the receiver will notice the signal. The sender may also select a site in which to produce the signal that has a simpler background or that is moving in a very different way.

Another major way to catch a receiver's visual attention is by increasing the contrast between a signal body part and the background. Black-and-white patterns are a common solution. The black is generated by using one of several synthesizable proteins called melanins. White is created by inserting small crystals or bubbles into the organ surface. They are large enough that they scatter all light wavelengths, creating white stimuli. Colours other than black and white are more difficult to produce. Although animals cannot synthesize carotenoid pigments, they can sequester the pigments by eating certain plant parts or by eating other animals that ingest those plants. Carotenoids are relatively stable compounds, generating colours in the yellow-to-red range. There are few natural pigments that are blue or green that animals can utilize for coloration. One exception is the chemical combination of carotenoids and proteins used by arthropods to colour their carapaces (hard outer coverings) dark green or blue. Crabs and lobsters turn red when cooked because the pigment proteins are denatured with heat, releasing the carotenoids.

Other animal groups use nonpigment techniques to produce blue or green coloration. One method uses a checkerboard matrix of alternating more-dense and less-dense materials in an external surface layer to selectively scatter certain wavelengths. The colour reflected depends on the size and spacing of the matrix. This mechanism is responsible for producing the blue and green feathers of jays and parrots, the blue skin on the heads of turkeys and male mandrills, and blue eyes in humans and snow leopards. Another technique is to use two thin layers of reflecting material on external surfaces. If the layers have the correct thickness, certain wavelengths are reflected by the two layers out of phase (the crest of one wave coincides with the valley of a second wave), thereby canceling each other out. The remaining light waves are in phase (the crests of the light waves coincide) and are visible as intense colours. Which light waves are canceled depends on the viewing angle of the receiver. Thus, the apparent colour can change as the sender shifts its position relative to the receiver. These are the iridescent colours seen in many hummingbirds and butterflies.

While it is often assumed that ambient light is white, every local environment is actually bathed in its own particular mixture of light frequencies. On a forest floor there may be only small amounts of red and blue light in the ambient mixture because green-reflecting plants absorb both red and blue light strongly. Fine particles in the atmosphere and the dense molecules in water both scatter the shorter blue and ultraviolet light frequencies but transmit the longer red and orange frequencies. The mixture of ambient-light frequencies can be a strong selective force favouring certain colours for signals. Where there is little ambient light, an animal may have to produce its own light (bioluminescence). This ability has evolved in fireflies, ostracod crustaceans (mussel shrimps), deep-sea squid, and marine fish, all of which produce their own light chemically or harbor luminescent symbionts that can produce light for them.

Fireflies such as those of the genus Photinus produce their own light as a means of communication.

A third major modality used by most animals (with the exception of many birds) is olfactory signaling. Senders deposit or release chemicals called pheromones that receivers later detect by smell or taste. The cost to senders of chemical communication can be minimal, as when feces or urine is used as a signal, or can be substantial, as when complex organic molecules must be synthesized solely for the purpose. Ants are prolific users of chemical communication. Their small bodies contain up to a dozen separate glands, each producing a different chemical compound or mixture of compounds that serves a different social function. Ants use chemical signals to mark their foraging trails, to recruit group members for defense against invaders, to attract mates, to regulate development of different worker castes, to solicit food, and to distinguish colony members from non-members. Many mammalian species employ chemical communication for important social functions, such as mate attraction, synchronization of mating, and territory defense. Mammalian glands often produce specialized chemical products, but some species mix various natural body products into a pouch and let bacteria do the work of producing the final pheromone product.

The potential for signal diversity is extremely high in chemical communication, as is the opportunity to create a signal that is very different from background odours. However, this diversity is often constrained by the degree to which chemical signals are appropriately volatile in air or soluble in water, resistant to degradation after release, and detectable by receivers. Animals often limit the volatility or solubility of pheromones by embedding them in an inert carrier compound that releases odorant slowly. This technique is particularly useful for territorial defense. Desert iguanas carry this one step farther by using a carrier that does not release odorant until another lizard flicks the carrier with its wet tongue. The small packets of deposited pheromone absorb ultraviolet light and thus appear as black specks to animals that can see in the ultraviolet spectrum, such as iguanas. Iguanas approach the specks and taste them, thereby releasing the pheromones. Other animals accelerate the dispersal of relatively non-volatile scents by spreading them over a tuft of fine hairs, actively spraying them into a medium (e.g., air or water), or releasing them into strongly flowing wind or water currents.

Tactile signals involve special patterns of touching, generating persistent eddies in a medium, or the transmission of vibrations through a substrate. Touching during aggressive encounters may provide information about the body size and strength of opponents. The grooming of another individual, called allopreening or allogrooming, has both hygienic and signal functions in many birds and mammals. Courtship signals may include a tactile component for synchronizing mating or gamete release. Roosting with body contact not only preserves heat but also appears to signal pair or group affiliations in mammals and birds. Arthropods make wide use of tactile, eddy, and substrate signals. For example, aquatic male copepods can identify the distinct eddies left by swimming females and track them for mating. The dances of honeybees are usually performed in a dark hive, and attending workers monitor the dancer with their antennae; some signal vibrations may also pass through the honeycomb substrate. Other arthropods attending to substrate-borne vibrations include water striders (using the surface of water), spiders (using their webs), and leafhoppers (using their host-plant stems and leaves).

Electrical discharges can also be used for signals. Two orders of freshwater fish, the Mormyriformes of Africa and the Gymnotiformes of the neotropics, are nocturnal or live in muddy water. These fish create an electric field and use distortions induced by nearby objects to navigate and find food. Not surprisingly, they also use the same electrical discharges to communicate with each other. In both groups, special bioelectric organs have evolved to produce rapid trains of discharges. The waveforms of these discharges vary with species and even with sex, and the rates of discharge can be modulated in complex ways to mediate social interactions. Because of the high resistance of fresh water and the low voltages the fish produce, electrical communication is limited to a distance of about 1 metre (3.3 feet).

Signal Transmission

All animal signals degrade as they propagate between sender and receiver. The farther apart the two parties, the greater this degradation will be and the less a signal will stand out from background noise. Senders can do little to reduce degradation once the signals have left the sender. However, they do have a choice of what kind of signal they produce, and evolution has often favoured choices that minimize degradation.

Elephants communicate by using low-frequency sound waves that can be detected several kilometres away from the animal sending the signal.

Sound signals transmit efficiently over large distances, around obstacles such as trees and foliage, and in dark environments. Nevertheless, sounds of all frequencies become less intense as they radiate away from a source. Higher frequencies suffer additional attenuation owing to heat losses and scattering of the sounds. Since small animals can produce only high frequencies (short wavelengths), their sound communication is often limited to short distances. Furthermore, ambient sound is often greatest at low and high frequencies, making intermediate frequencies the ones least likely to be obscured by the background.

Propagation of sound is complicated when the sender and receiver are close to a boundary (e.g., the ground for terrestrial animals and the water's surface for aquatic animals). This is because sound can travel to a receiver by two routes: a direct route along the line connecting sender and receiver, and an indirect route in which the sound bounces off the boundary and up to the receiver. If the two replicates of the sound signal arrive at the receiver out of phase, they will cancel each other out. The closer both parties are to the surface, the more acute the cancellation. Low frequencies are most susceptible to these effects. However, for terrestrial animals, very low frequencies can propagate by a third route, called a ground wave, if the surface is sufficiently porous. Intermediate frequencies, however, are still canceled out even when the ground is porous. As a result, intermediately sized animals that cannot produce low-frequency sounds often climb or fly to locations high above the ground before vocalizing. Elephants, which cannot fly or climb, resort to sufficiently low frequencies that they can be detected several kilometres away. Whales also produce low frequencies and move sufficiently far beneath the ocean surface before vocalizing, which enables their signals to be heard hundreds of kilometres away.

The optimal temporal patterning for sound signals also varies with habitat. Rapid temporal patterns quickly become unrecognizable owing to echoes in heavily forested habitats. In contrast, sounds propagating in open grasslands suffer little from echoes but instead acquire slow artifactual modulations because of air turbulence. Birds, even of the same species, are much more likely to use rapid temporal modulations of their calls, such as trills and buzzes, when they live in grasslands than when they live in forests. Forest birds typically produce long whistlelike notes with slow, if any, modulations.

Whales such as humpbacks (Megaptera novaeangliae) communicate by producing low-frequency sound waves. The animals move sufficiently far beneath the ocean surface before vocalizing, which enables their signals to be heard over hundreds of kilometres.

Light signals also suffer transmission losses. Intervening obstacles such as foliage easily block the straight-line propagation of visual signals, and increasingly distant senders occupy a decreasing part of a receiver's visual field. Light waves are also subject to filtering and scatter that can distort a signal pattern and decrease its contrast with the background. All of these effects make detection and recognition of a visual signal more difficult at a distance. In addition, reflected light signals require some source of ambient light, and visual communication thus becomes more difficult to achieve at night and in very dark environments. Bioluminescence is, of course, one solution to this problem.

Water jelly (Aequorea victoria).

The medusa stage of the bioluminescent water jelly or crystal jelly (Aequorea victoria) in the waters off the west coast of North America. The species is harvested for its luminescent protein called aequorin, which has been used in medical research as a fluorescent marker protein.

Olfactory signaling differs from sound and light communication in significant ways. Pheromones spread from a source by diffusion and medium turbulence. This process is much slower than the propagation of light or sound signals, and its erratic path can make it difficult for a receiver to locate an odorant source. Whereas sound and light largely retain their temporal patterning as they propagate, temporal patterning of pheromone release is quickly lost because of turbulence. The slow speed, the limited ability to be located, and the loss of temporal pattern constrain the uses of olfactory communication to short-range signals and to recurrent functions, such as territory defense and mate attraction.

The unpredictable effects of turbulence also limit the range over which an olfactory signal can be detected. Receiver moths get around this constraint by sampling wind direction and flying into the wind to trace the chemical source (a process called anemotaxis). Many nocturnal mammals deposit scent marks at multiple locations within their territories. Although no single scent mark is detectable at a distance, the ensemble ensures that any intruder is aware of the owner's presence in the territory. Some insects and marine species actively seek out sites where air or water currents can convey olfactory signals over long distances.

Signal degradation during propagation is not always a detriment. Many birds use degradation of songs to estimate the distance to competitors. As long as there is sufficient degradation, a territory owner can conclude that the other singers are too distant to be threatening. Many social insects use olfactory signals to mark feeding trails, to give alarms, or to identify colony mates. To avoid confusion and focus attention on a specific location or individual, it is best if these signals have only limited ranges of detection.

Regardless of the modality used, senders communicating with distant receivers face a number of competing influences. Body size, habitat type, time of day, proximity to a surface, and speed with which a receiver must respond all affect the form of the optimal signal. Because species differ in these factors, optimal signals differ between species even when communicating the same information. The constraints imposed by the physics of signal production and transmission account for an important fraction of the diversity seen in animal signals.

Signal Reception

The receiver's task is to detect signals against the background and to discriminate between different signals. Most animals use the same sense organs (eyes, ears, noses, touch receptors, etc.) for signals that they use to detect other external stimuli. Their brains also process all sensory stimuli, both signals and non-signals, with similar procedures. However, communication is sufficiently important that most animals show some tuning of their sense organs and some specialization of their brains to improve signal detection and characterization.

Crickets have eardrums on their legs that are connected by tubes to breathing holes on the sides of their bodies. This allows them to obtain multiple samples of a given sound field at each point in time.

Detection of sound is often challenging because the received signals are faint and distorted owing to propagation. Sound traveling in air is largely reflected from solid objects, including animals, with little energy transfer. Sound traveling in water is easily transferred to aquatic animals, but because all parts of the animal vibrate in synchrony, there is no immobile reference allowing the animal to detect the vibrations. As a result, animals have had to acquire some very sophisticated adaptations to hear sounds.

Terrestrial animals often have funnel-shaped structures outside the body to collect and concentrate impinging sounds. The funnel shape also creates a gradual change in the properties of the sound-propagating medium from that of air to that of liquid and solid bodies. This increases the amount of trapped sound energy. At the end of the funnel is a thin membrane (called an eardrum) that is set into vibration by the sounds. Small bones or fibres transfer the eardrum movements to a fluid-filled cavity, within which are sensory cells bearing hair-like cilia. As the fluids move, the cilia sway, thereby stimulating the attached nerve fibres. Ears of orthopteran insects (e.g., crickets and grasshoppers) and vertebrates are designed such that different frequencies of sound stimulate different sets of sensory cells. Thus, the animal is able to decompose a complex sound into its component frequencies. This enables the receiver to separate signal frequencies from those of non-signal sounds. For each band of useful frequencies, amplitudes and a temporal pattern are encoded in the rates and patterns of nerve impulses sent to the brain. Larger animals usually have two ears and use the time delays or differences in signal amplitude at the two ears to identify the direction of the sender. Small animals cannot use this method because the delays and intensity differences are very small. However, they can achieve some directional information by sampling a sound field at several points simultaneously. For example, crickets have their eardrums on their legs, and fine tubes connect each eardrum internally to breathing holes on the sides of their bodies. This allows them to obtain multiple samples of a given sound field at each point in time. Frogs and birds have a space connecting their eardrums that works in a similar manner.

Some aquatic animals have exterior cilia or hairs that sway as sounds pass over them and stimulate sensory cells. This mechanism is effective only within a few wavelengths of the sound source and tends to be limited to lower frequencies. For sound reception at a greater distance from the source, an aquatic animal must create body parts that are moved in different ways during the passage of sound waves. Some bony fish use swim bladders for this purpose. These are air-filled sacs that provide buoyancy. Because air is much more compressible than water, sound trapped by the swim bladder results in much larger molecular displacements than the same sound energy propagating in the rest of the fish's body. Connecting the swim bladder to the fish's ears makes it possible for the differences in molecular displacement between the swim bladder and the rest of the fish to be used to stimulate the sensory cells of the ear. This allows a fish far from the sender to detect and measure the passing sound waves.

Animal eyes differ markedly in their range of view, their resolution, and their focusing power. The eyes of vertebrates and cephalopods (octopus and squid) have adjustable lenses that extend the range over which images are in focus. They also have an iris that adjusts the amount of light entering the eye. An effective eye must have many receptor cells if it is to preserve the relative positions of the different objects reflecting light to it. To discriminate between different colours, different receptor cells must be sensitive to different light frequencies. In addition, because light intensities are highly variable, eyes may need different receptor cell types to handle both dim and bright light. Vertebrate eyes use cone receptor cells for bright light and colour discrimination and colour-insensitive rod receptors for dim light conditions. All light receptors contain a protein-bound pigment. This pigment, called rhodopsin, occurs in all multicellular animals. These pigment molecules change shape when absorbing light. This triggers a chain of reactions within the receptor cells ending in the production of nerve impulses. Different photopsins and rhodopsins absorb different light frequencies, permitting receptors to differ in colour sensitivity. Arthropod eyes consist of 8–10 receptor cells clustered around each of many facets, or corneal lenses. It is the number of facets, not the number of receptor cells that determines visual acuity in arthropod eyes.

Unlike the case with auditory signals, the direction of a visual sender is automatically known once a receiver has detected the signal. The relative positions of receptor cells in an eye are preserved in their projections to the brain. This allows the brain to create a map that replicates, to varying degrees, the visual field of the eye. Projections from the visual maps of the two eyes are compared to identify specific objects and their relative distances. The recognition of patterns is a function of the brain and relies on a combination of inherited and learned mechanisms.

Pheromone reception is accomplished by smell (olfactory) or taste (gustatory) organs. These receptors contain sensory cells with fine cilia, or microvilli that extend into the medium. Pheromone molecules and other stimuli temporarily bind to specific proteins on the cilia or microvilli. The binding triggers a cascade of chemical reactions within the sensory cell that result in the production of nerve impulses flowing to the brain. The more molecules that bind to a cell, the higher the rate of nerve firing. The sensory cells in an olfactory organ can have highly specific or generalized sensitivities. Specific cells tend to respond only to a certain molecule. Animals often use such cells to detect species-specific mating pheromones. Generalized cells respond to several types of molecules. They are used for less-specific functions, such as the recognition of food items. The olfactory organs of different species vary in the mix of specialized versus generalized sensory cells that they host. Olfactory organs can be as simple as a patch in a mammalian breathing passage or as complex as the plumose antennae of male moths. In some mammals the vomeronasal organ (Jacobson's organ), located in the roof of the mouth, is used to mediate a behavioral response known as flehmen, in which an animal raises its head and lifts its upper lip in reaction to specific odours. This response requires special movements of the tongue and lips to admit chemical samples to the sensory cells. The vomeronasal organ is the primary receptor organ for many of the pheromones that dictate interactions in mammalian social life, including pheromones involved in conflict, reproduction, and parental care.

Red imported fire ants (Solenopsis invicta), as well as other ant species, create marker trails using pheromones. A pheromone trail is followed by worker ants traveling between the nest and a food source.

Electroreception appears to have been a widespread sensory ability in primitive fishes, in which it was used to detect muscle and nerve impulses in hidden prey. Modern sharks and rays still use this technique for hunting. Mormyriform and gymnotiform electric fish developed these primitive receptors into sophisticated tuberous organs that are used in social communication. These structures, embedded in the fishes' skin, are encapsulated in ways that make them insensitive to slowly varying electric fields, such as those produced by muscles, but responsive to the rapid discharges of other electric fish. Tuberous receptors are usually tuned to be most responsive to the discharge rate of their own species. The brains of these fish are also highly sensitive to changes in the repetition rates of discharges. This permits sophisticated exchanges between fish during conflicts, courtship, and territorial defense.

Sharks such as hammerheads have sensitive structures called electroreceptors embedded in their skin. These structures are responsive to electric discharges produced by other fish and serve an important role in communication.

Costs and Benefits of Communication

For both senders and receivers there are costs associated with engaging in communication. It takes time, energy, and special modifications of sender and receiver organs to communicate. Thus, there must be compensatory benefits to each party for communication to be favoured by evolution. A sender will provide information to a receiver only if the decision of the receiver improves the sender's fitness more than the costs of signaling reduces it. The benefits to the sender may be direct, such as securing a mate or successfully repelling an opponent, or indirect, in that the receiver's choice may benefit close kin of the sender. A receiver attends to any source of information that is sufficiently reliable, on average, to enhance the receiver's decision making. The qualifiers "sufficiently reliable" and "on average" reflect the fact that senders may not always send perfect information, but the signals may still be useful to receivers. However, there is a minimum amount of reliable information in any decision situation that must be provided before it is beneficial for either party to engage in communication. Described another way, the payoff to each party (benefits minus costs) must be positive before they will participate in communication.

Animals often have to make decisions in response to alternatives. For example, if a female must determine which of two males will be the better parent for her young, she will mate with the male she deems most fit. In another example, when a parent bird returns to its nest, it must decide which of its nestlings is most needy. It will then give that nestling the worm. The receiver's decision process begins with some baseline probabilities for each alternative, based on inherited biases, chance, or prior experience. As new information is obtained, whether by examining the candidates directly or by attending to their signals, the receiver updates the probabilities by raising some and lowering others. The receiver's decision will then be based on these updated probabilities and some knowledge of the relative payoffs of correct versus incorrect decisions.

A number of measures have been proposed for the amount of information provided by animal signals. One measure is relative reliability; this is the increase in the average probability that a receiver makes the correct decision (for itself) when using signals compared to when a receiver is not using signals. Relative reliability also may be weighted according to the relative chances of making the correct decision; this is done by determining the difference in payoffs when making the correct versus incorrect decision. A classical approach computes a logarithmic measure (in bits) of the reduction in a receiver's uncertainty after receiving and processing a signal. Most of these measures yield similar rankings of the amount of information in signals and differ largely in scale.

Few senders never err in signal selection, and even if they selected the correct signal, transmission can distort signals and cause receivers to interpret them incorrectly. Thus, most animal signaling systems have reliabilities less than that expected for perfect information. Receivers usually have a "best guess" that they adopt when they cannot obtain enough information from a signal to make an informed choice. This best guess is the one that yields the highest average payoff when no signals are available. A receiver that switches from best guesses to a reliance on imperfect signals will start making incorrect decisions, when the best guess was in fact the correct choice. However, the receiver will also start making some correct decisions on those occasions when the best guess was the incorrect choice.

For receivers to participate in communication, the benefits of not making errors when the best guess was wrong must be greater than the costs of occasional errors when the best guess was always correct. As reliability increases, these benefits keep growing and the costs become diminishingly small. Thus, there is a minimal reliability that is required before either party should engage in communication.

Evolution of Signals

New signals do not evolve from scratch. As with any adaptation, new signals evolve from existing body structures, organs, physiological processes, and ordinary behaviours that animals already possess for non-signaling functions. These are sometimes called protosignals. Since the sender can benefit only when the receiver can interpret the protosignal, the receiver must already possess some ability to detect it. Thus, both parties must have prior adaptations that already facilitate the exchange of information. The sender's protosignal may have been initially poorly associated with the context of interest to the receiver, and the receiver's reception organ may not have been very effective at detecting the protosignal. However, once such precursors are in place, each party can take advantage of the other, and this can be sufficient to initiate subsequent coevolution of both signal generators and receptors.

Historical scenarios for signal evolution fall into two categories. Scenarios emphasizing sender precursors were a major focus during the early days of ethology in the 1950s and '60s. The Austrian zoologist Konrad Lorenz, who founded the field of ethology, noticed that the courtship displays of many birds appeared to be elaborated versions of simple preening movements, feeding actions, or nest-building activities. Dutch zoologist Nikolaas Tinbergen, as well as other scientists, provided many subsequent examples of the similarity between mate attraction displays and ordinary survival behaviours. In this scenario, receivers begin to notice actions of other animals because these provide cues about what the other animals will do next. Thus, the protosignals exist prior to receivers' noticing them. The second perspective, called sensory drive, emphasized receiver precursors for the evolution of signals and was developed during the 1980s and '90s, spearheaded largely by American biologist John Endler. In sensory drive, signals were viewed as new behaviours or structures that exploit existing sensory biases of receivers. For example, existing female search behaviours for particular foods or offspring might be mimicked by males to get the attention of females for courting. With new DNA (deoxyribonucleic acid) technology for generating accurate evolutionary trees, scientists have been able to trace the histories of signals and receptors using one or both of these scenarios.

Whichever route creates the association between protosignal and context, the subsequent coevolution of signaling and receiving organs tends to follow a similar trajectory. The protosignal often undergoes a reduction in the number of components and an exaggeration of the remaining components in a process called ritualization. Ritualization of visual signals often involves the addition of colour; elongated or erected fur, feathers, and fins; or enlarged body structures that enhance the visibility of the display. As a result of ritualization, auditory signals may be shortened, repeated, or modulated in various ways that make them distinct against a noisy background, and chemical signals are enhanced with structures and behaviours that maximize odour dissemination. The receiver organ may then be fine-tuned to make it especially sensitive to the critical signal components emphasized during ritualization.

Signals that evolve from cues via the sender precursor route are associated with a particular context or meaning from the outset. This evolutionary process is demonstrated by the predator alert signal of Thomson's gazelles (Gazella thomsoni). A gazelle that has heard a suspicious sound is likely to stop foraging, raise its head high, and stare in the direction of the sound with ears pointed forward. Nearby individuals, spotting the frozen head-high posture, gaze in the same direction and prepare to flee. The freeze reflex has become ritualized and amplified through the evolution of black-and-white stripes along the gazelle's face and body that make the staring posture more conspicuous.

In figure, thomson's gazelles (Gazella thomsoni) use a ritualized alert signal to communicate the presence of a potential predator. This signal is characterized by a frozen posture in which the head is held high in the air and is pointed in the direction of the threat. Nearby individuals interpret this behaviour as a sign to prepare to flee.

Intention movements are another widely cited source of signals. They consist of incomplete or preparatory acts that precede major social activities, such as attacking, mating, reconciling, or fleeing. For example, baring teeth, horns, and claws, staring forward, and covering sensitive body parts such as the ears are some well-known precursors for subsequent attack behaviour. Aggressive individuals have ritualized many of these intention movements into signals when both parties benefit by avoiding costly fights. In dogs and cats, the display is ritualized by the exaggeration of lip and nose wrinkling and by the addition of a vocal component. Submissive signals are often the precise opposite in form from a species' aggressive signals and may consist of an averted gaze or a closed mouth. This contrast between aggressive and submissive signals is called the principle of antithesis and was first noted by British naturalist Charles Darwin in 1872. Other intention movements that have evolved into signals with clear information about what the sender is likely to do next include mating postures during courtship, presentation of nest material to stimulate breeding, outstretched arms to indicate friendly intentions, and head flips by ducks that coordinate taking flight by mimicking the physical act of jumping into the air.

A long-eared owl (Asio otus), its feathers spread in an aggressive display.

When animals are thwarted in achieving a goal or simultaneously experience two conflicting motivations, such as fear and aggression, they sometimes perform brief, irrelevant behaviours, such as mock feeding, sleeping, or preening. These are called displacement behaviours, and they sometimes become ritualized into displays that appear to indicate the ambivalent state of the sender. In other cases of conflicting motivations, animals may blend two antithetical displays. For example, the broadside threat display often seen in the early stages of a conflict between two competitors in animals such as ungulates, cats, lizards, and fish has been viewed as the blending of aggressive approach and flight. In these animals, the broadside display also presents the largest possible body profile to the opponent, and the display may be thus enhanced with erected fur, feathers, or fins. Aggressive individuals that are fearful of their opponents may also perform displays of redirected aggressive attacks on nearby inanimate objects, reminiscent of an angry person who slams a door instead of causing physical harm to the individual who is serving as the source of frustration. The form of vocal signals can also reveal information about sender state.

Vocalizations associated with fear tend to be high-frequency screams because the fearful individual reflexively tenses all of its muscles. This increased muscle tension stretches the membranes used in sound production, causing them to vibrate at higher frequencies. On the other hand, aggressive intent is expressed with low-frequency vocalizations because the threatening individual is more confident and because low frequency is associated with larger body size and thus is inherently more threatening. Some chemical signals are derived from steroid hormones and are used to inform receivers about the sex and reproductive state of the sender. New displays sometimes evolve from other displays in the species' signal repertoire. For example, a female primate's copulation solicitation display of rear-end presentation is used in modified form by subordinate males to reduce aggression by dominant males.

Most examples of signal evolution via the receiver precursor scenario involve mate attraction. Tropical anole lizards possess specialized motion detectors in their eyes for distinguishing the jerky motion of their insect prey against the waving motion of the background foliage. The push-up displays of males that are used both to attract females and to repel other males mimic this jerky movement and thus ensure the visibility of their displays. This behaviour is sometimes coordinated with dewlap signaling, in which a male displays a large, colourful throat fan. Many nocturnal moth species have evolved specialized ears for detecting the high-frequency calls of their bat predators. Mate attraction in these moth species is achieved with an olfactory signal, usually given by the female. One group of moths has become diurnal (active during the day) and thus no longer suffers from bat predation. Males in these species have taken advantage of the females' auditory biases and thus have evolved the ability to produce a high-frequency sound to attract mates. In water mites, tactile animals that do not have eyes or ears, males attract mates by vibrating the water surface with their forelegs—a behaviour that mimics their insect prey. Male mate-attraction signals that evolve via this process are highly species-specific because they are tuned to the receptors of the females of their species.

Signal Design Rules

The diversity of animal communication signals is enormous. Each signal is assumed to have converged on the form that is best adapted to transmitting the type of information conveyed by that signal in a given social and environmental context. The form of a signal is therefore affected by its modality, the habitat through which it must be transmitted, and the social function it serves. This fact first precipitated a search for general principles of signal design in the 1960s, when American linguist Charles Hockett and American behavioral biologists Stuart Altmann and Peter Marler attempted to come up with lists of design features that could then be used to characterize any type of signal. Once the design feature requirements, or rules, for a given type of signal had been established, the specific mechanisms for satisfying the rules in each modality could be used to determine whether signals serving the same social function in different species exhibited the expected similarities in signal form.

A minimum list of six design features seems to be sufficient to characterize most signals. The range of a signal is the distance that the signal must be transmitted, which is in turn determined by the typical distance between sender and receiver for the social context in which the signal is given. Signal range can be optimized by adjusting amplitude, intensity, hue, size, concentration, and signaling location, depending on the modality. The ability of a signal to be located specifies the degree to which the location of the sender, the receiver, or some external object needs to be encoded in the signal. The form of the signal and the directional capabilities of the receiver's receptor determine how easily the sender's position can be estimated, and some modalities have better mechanisms for pointing toward a referent than other modalities.

The duty cycle of a signal is the relative amount of time that the signal needs to be on versus off. Some types of signals require on times that are much longer relative to other signals, and the duty cycle can be optimized by adjusting signal duration and repetition rate. The identification level of a signal determines the number of different units that must be distinguished by unique signal variants. Identification levels, listed in order of increasing number of variants required, include species, sex, age groups, colonies, family groups, and individuals. The larger the number of variants the signal must encode, the more complex the signal form must be to generate these variants. In contrast to this between-individual level of variation, the modulation level specifies the degree to which the signal must vary within an individual. Some signals need to encode graded levels of information, whereas other signals need to encode discrete, presence-or-absence information. In addition, the form-content linkage specifies the degree to which the form of the signal is linked to its information content. Signals that indicate what the sender is likely to do next are limited to forms that are functionally linked with the subsequent action, whereas signals that indicate location and identity can have more arbitrary forms.

The design rule for each of these features depends on the signal's social function. For example, mate-attraction signals are usually given by one sex for the purpose of attracting members of the opposite sex. The information encoded in the signal includes species identity, reproductive maturity, and sender location. Therefore, such signals must be designed to have a large range, good sender-locating ability, high duty cycle, species specificity, and low within-individual modulation. Signals can be arbitrary in form if they transmit only species identity and location. However, if the nonsignaling, choosy sex must select from several signaling suitors, it may seek information about sender quality, health, parental abilities, or other attributes, and the signal will evolve to contain detectable characteristics that are linked to these qualities.

Territory-defense signals inform other nearby individuals that an area is occupied by a defending owner. Territorial signals have many of the same design rules as mate-attraction signals, including long range, high duty cycle, and species specificity. They also can be arbitrary in form, unless they encode information about the vigour or fighting ability of the owner. In order to establish stable boundaries, owners must be able to identify their territorial neighbours, and territory defense signals need to encode individual identity. Territory boundaries may be as important to locate as the territorial owner, a function that long-lasting olfactory marks deposited around the periphery can achieve very efficiently.

Courtship signals are typically given by males once a male and female have approached one another. These signals function to persuade the female to mate, since females are usually more choosy and reluctant than are males. These signals also serve to coordinate the mating act. The range of a courtship signal should be small not only because the sender and receiver are close but also because the mating couple does not want to attract interlopers or predators. Therefore, in most cases, sounds, movements, and scents are low in amplitude. Sender location is generally irrelevant, but in species that lay eggs in specific types of sites, the male may need to point or direct the female to a location he has found or prepared. Duty cycle is typically high for a brief period, and the repetition rate of a signal may increase gradually in order to synchronize copulation, or gamete release in externally fertilizing species. Signals need to be species-specific, and they often need to be sex-specific as well. There may be some need for within-individual modulation if either of the sexes encode their motivation to proceed with the mating. In contrast to fighting or other activities, males often signal their intentions to mate by evolving displays linked to the mating act, and females signal their willingness to mate by assuming a posture that facilitates mating. Courtship generally involves several different signals in multiple modalities, with tactile and olfactory signals often playing an important role. Signal form varies widely, depending on the details of the reproductive biology and habitat of the species.

Raggiana bird-of-paradise. A male Raggiana bird-of-paradise (Paradisaea raggiana) displaying its colourful feathers during courtship.

Threat signals are given when two individuals compete directly and at close range over a non-sharable resource, such as food, a mate, or a territory. With the exception of ritualized fighting, reciprocal communication is used in an attempt to resolve conflicts without fighting; therefore, both individuals are simultaneously senders and receivers. The conflict is likely to be won by the individual that is larger, stronger, healthier, more experienced, or more motivated. Threat signals are designed to transmit information about these sender qualities. Once one contestant decides it would lose during further escalation, it gives a surrender signal to end the conflict. Threat signals share some design features with courtship signals: their range is short; displays are directed at specific individuals; duty cycle is high for a brief period, modulations that encode gradations in motivation are required; and signals are often linked to intentions. However, there are some significant differences. Signals are usually short, forceful, and conspicuous. Identification level involves the recognition of rival status, which may be based on discrete age or sex classes or on a continuous range of classes based on dominance rank or body size.

Modulation of threat signals may transmit information about aspects of fighting ability that vary within individuals, such as current condition and motivation. Threat signals are often ritualized intention movements, ambivalent combinations of acts, or redirected behaviours, and some are linked to the size or health of the animal. As with courtship, there are often numerous threat signals in a species' repertoire, with visual, vocal, and tactile signals providing redundant information about relative motivation and fighting ability.

A final type of signal, one that is observed only in relatively social species, is the alarm signal, which provides information about the presence of a predator or a conspecific rival. There are two different kinds of alarm signals: Flee alarms, given in the context of a cluster of animals in immediate danger that cause receivers to rapidly disperse and hide, and assembly alarms, given in the context of dispersed animals that cause receivers to move toward the sender for some type of joint rescue or mobbing response. Design rules are quite different for these two signals. Flee alarms, which are not especially loud because receivers are nearby, are designed to prevent the sender from being located (especially by the predator) and may be linked in form to fearful internal states. Assembly alarms, which have a larger range and longer duty cycle than flee alarms, require that the sender be locatable. The signal is usually repeated and modulated to indicate the degree of urgency. The first real evidence for the validity of design rules was demonstrated with these two types of alarm signals. Many small birds possess vocal alarm signals—one that is directed toward hawks flying overhead and another that is directed toward owls and ground predators, which are often mobbed. Hawk alarms show a striking convergence on a form consisting of a single relatively long high-frequency whistle that has a gradual onset and gradual offset. This is a very difficult sound for a hawk to detect and localize. In contrast, owl alarms consist of short repeated broadband notes that are easy to locate by predator and prey alike.

Honesty and Deceit

Senders and receivers may have conflicting interests in the accurate exchange of information. Among humans, it is known that exaggerating and lying can sometimes benefit senders. Animal senders may also gain fitness by cheating under certain circumstances; the strength of the selective pressure to do so depends upon the signaling context and the degree to which the two parties have conflicts of interest. Conflict of interest is greatest when two more or less equal competitors both desire the same non-sharable resource. Each would like the other to back down without a fight, and each would benefit from persuading the other that it is the better fighter by any means possible, including bluffing. In the mate-attraction context, both male and female benefit from mating with the correct species and therefore agree about the accurate transmission of species information. But females may want to mate only with a high-quality male, which puts pressure on low-quality males to hide or exaggerate their quality. An offspring in a multiple brood may exaggerate its need for food to the parent in order to garner a larger share of the food for itself.

A chick in a group of nestlings may exaggerate its need for food in order to gain a larger share of food for itself.

The problem of signal honesty is an important issue in studies of animal communication systems. In the early days of ethology, signals were shown to evolve through the ritualization of behaviours that are, or were, functionally appropriate to the contexts in which the signals are given. Signals were believed to be honest indicators of underlying motivations because the signals were derived from physiologically or anatomically linked sources. With the rise of evolutionary game theory in the 1970s, this notion of signal honesty was questioned. British ethologist and author Richard Dawkins and British zoologist John Krebs suggested that senders were best characterized as deceitful manipulators trying to mask their true intentions and trick receivers into actions benefiting senders. Thus, receivers were best viewed as mind readers trying to discount false signals, anticipate the true intent of the sender, and identify their own best countermove. This scenario leads to a never-ending arms race with increasing deceit and concealment of true intentions by senders parried by increased discrimination and exploitation by receivers. Except where sender and receiver have common interests, the resulting signals are largely deceitful and uninformative.

Israeli evolutionary biologist Amotz Zahavi challenged this pessimistic view of signal honesty. He asserted that receivers have the upper hand and should not respond to signals unless they carry some guarantee of honesty. One guarantee is to require that signals impose a cost such that deceitful senders cannot afford to produce an exaggerated signal, or they produce it only in an ineffective way. Signals characterized by such costs are called handicap signals. Although Zahavi's idea was viewed skeptically at first, subsequent game theory models demonstrated the evolutionary feasibility of handicap signaling, and hence the handicap principle became widely accepted.

A key concept of the handicap principle is that the cost imposed by the signal must be closely related to the sender quality attribute about which the receiver wants information. Although Zahavi proposed that the signal should be designed to "use up" the sender's quality feature in a display of costly consumption, it is more important that a sender of lower quality with respect to the attribute be less able to afford to produce the signal. Thus, the form of the signal is linked to its information content. One type of cost is energy expenditure. Signals with high production costs can inform receivers about the health, vigour, or foraging abilities of senders. For example, vocal and visual mate-attraction signals must be repeated again and again. Female preference for males that not only produce high-quality displays but also repeat these displays at a high rate will increase the selection pressure on males to perform at the highest possible energetic level they can sustain. Unhealthy or poor-quality males cannot maintain such an expensive display, and this fact will be detectable to females. Choosy females benefit by acquiring vigorous mates that are good genetic fathers and parental providers or possess food-rich territories. Handicap signals also may be given to potential predators. For example, antelopes sometimes perform energetic jumping to approaching predators. Only individuals in good condition can perform these actions well. This provides honest information to the predators that discourage them from chasing the able displayers.

Handicapping is not the only mechanism for generating honest signals. Some signals are constrained by physiology, anatomy, or physical principles to be honest indicators of certain types of sender attributes. Such signals are called unbluffable, or index, signals. Examples of index signals include low-frequency vocal threats that are linked to body size in many vertebrates and tail-beating displays in fish. Likewise, certain forms of ritualized fighting, such as mouth wrestling and antler locking, are index signals of weight and strength. Other signals are associated with aging and health, including song repertoire in birds, which can indicate age, experience, and ability to survive. In addition, pointing displays that indicate the direction of gaze and olfactory signals that are related to reproductive physiology are considered index signals.

Rival European red deer stags (Cervus elaphus) engaging in ritualized fighting for possession of a hind in the rutting season.

True threat signals of intention to attack require very effective honesty guarantees if they are to convince rivals to retreat. Otherwise, an initially honest signal will be invaded by bluffers that give the signal but never follow through with an attack. In this case the sender must demonstrate its sincerity by approaching the rival very closely. Only a threat display performed within close proximity to the rival, where there is the risk of a retaliatory attack, will be taken seriously by the receiver. Threat displays are typically attack-preparatory postures performed close to the rival, giving the sender both a tactical advantage and demonstrating its willingness to take risks.

Some communication signals are not costly to produce, risky to execute, or obligatorily linked with physical properties of senders. The code by which these signals are associated with contexts is an arbitrary convention, and these are therefore called conventional signals. If there is no conflict of interest between sender and receiver, senders will not be tempted to cheat; conventional signals can be honest and stable without further guarantees. However, conventional signals also are seen during conflicts of interest, and in this case there must be a stabilizing cost that maintains honesty. Conspicuous colour patches in some birds and lizards are the classic example of this type of signal. The size or hue of the patch is correlated with the dominance rank of the individual, hence the designation of these patches as badges of status. Large badge size deters aggressive challenges by small-badged individuals. The cost of guaranteeing honesty of a large badge is aggressive retaliation from other large-badged individuals. The evolution of such signals must be accompanied by frequent testing of the honesty of other individuals with a badge size similar to one's own while avoiding or ignoring individuals with larger or smaller badges. Such a rule makes it very dangerous and costly for a low-status individual to cheat by sporting a large badge.

The size of the black throat patch of the male house sparrow (Passer domesticus) is correlated with social status. Large patches are associated with high rank and good fighting ability.

Most signals are believed to be honest most of the time because conflicts of interest are minimal or because appropriate costs are imposed on cheaters. The best demonstrations of honest signaling have been described for handicap signals of mate quality. One well-documented example is the elongated tail feathers of the barn swallow (Hirundo rustica), studied by Danish zoologist Anders Møller. Females prefer males with longer tails, pairing very quickly with males having artificially enlarged tails compared with males with shortened tails. Long tails are a handicap for males. Barn swallows are aerial foragers that capture flying insects on the wing, and artificial tail elongation increases the drag on the tail and reduces agility and foraging efficiency. Males with naturally long tails are stronger, healthier, and resistant to parasites. These individuals not only grow long tails and cope with the foraging handicap but also transmit parasite resistance to their genetic offspring. Females obtain better-quality offspring by selecting long-tailed males. Other examples of honest mate-quality signals preferred by females include red plumage coloration in the male house finch (Carpodacus mexicanus), which is correlated with a male's foraging skill and survivorship; long call duration in the gray tree frog (Hyla versicolor), which is energetically costly for males but associated with better survivorship in their offspring; and high display rate in damselfish (Stegastes partitus), which is correlated with the survivorship of the eggs that the male tends.

In figure, the elongated tail feathers of the barn swallow (Hirundo rustica) are an example of honest handicap signaling. Long tails are a handicap for males because they increase drag and reduce agility. However, males with long tails tend to be strong and healthy and thus are the preferred mates of females.

Dishonest signaling does occur. Deceit is the provision of inaccurate information by the sender such that the sender benefits from the interaction but the receiver pays the cost of a wrong decision. Types of deceit include lies (using the wrong signal among an unordered set of alternatives), exaggeration or bluff (using a signal whose rank among ordered alternatives is different from that for the corresponding condition values), and withholding information (not giving a signal when appropriate). There are numerous examples of predatory species that mimic the mate-attraction signal of their prey, but in interactions between two different species; there is no selection pressure on the predator to be honest, and there is little the prey can do to avoid being exploited. The prey, as receiver, can try to improve its discrimination between true mates and impostors, but this process will simultaneously select for better mimicry by the predator.

Within-species deceit is a different matter, since dishonest signaling can sometimes backfire on the sender. For example, an outright lie has been described in birds foraging in flocks, where one individual may give a false alarm call to scare competitors away from a rich food find. A sender cannot "cry wolf" too often, however, because receivers may learn to ignore the signal, and the signal will thus cease to be effective in true alarm contexts. Bluffing threats sometimes occur in mantis shrimp (Gonodactylus bredini), which defend their burrows from intruders with a claw-spreading display. Recently moulted individuals that are soft and unable to defend themselves effectively may nevertheless sometimes give the threat display in the hope that an intruder will not press an attack. Withholding information has been described in primates that fail to advertise a rich food find. If other group members catch an individual feeding on such a find, the individual is aggressively punished. Thus, low levels of dishonesty may persist in many signaling systems. However, signals must be sufficiently reliable and honest most of the time; otherwise, they will be discounted and ignored by receivers, senders will no longer benefit from giving dishonest signals, and the signals will disappear from the species' repertoire.

Adaptations in Animals

An adaptation is something about an animal that makes it possible for it to live in a particular place and in a particular way. It may be a physical adaptation, like the size or shape of the animal's body, or the way in which its body works. Or it may be the way the animal behaves. Each adaptation has been produced by evolution.

As the environment changes, animals that cannot adapt die out, and only the adapted ones survive to produce babies. Because babies are usually more or less like their parents, the whole species soon contains only animals that are adapted to the new environment.

An animal's environment consists of many different things. The climate is important. Whether it is hot, cold, dry, or wet will have an effect on all the creatures that live in a particular place.

Another important part of an animal's environment is what kinds of food plants grow in it. The other animals that live there also have an effect. If there are predators around, the prey animals will have to learn to defend themselves or run fast to escape.

These adaptations make it possible for a great variety of creatures to live and thrive on earth. Animals adapt to the natural world.

Animals in the wild can only live in places they are adapted to. They must have the right kind of habitat where they can find the food and space they need.

Types of Adaptation in Animals

Some animals are much better able to adapt to their environment than others, but no species has survived intact. This is because the evolution of life on earth is a process. More precisely, it is a series of processes. Animals will adapt and change until they barely resemble their forebears. The first bird had to sprout wings, the first mammal had lungs, etc. There are limits to the ability of science to trace back any organism, but it is believed that some species such as jellyfish have been able to last millions of years. However, the likelihood of any animal this old being exactly the same is very small.

It is likely that all life on earth originated in a single cell organism. This means all life is paradoxical. No animal alive today existed at the beginning of life on the planet. Yet all life that exists today existed in this single cell. The different form this life takes is due to adaptation.

Although all animals are intrinsically adaptable, much of the specific adaptations happen for indeterminable reasons. Certain characteristics or behavioral traits have happened due to random mutations the genome. This doesn't mean that all failure of adaptation is due to genetics. Events such as climate change or natural catastrophes have resulted in the complete disappearance of many species on this planet. Although we are limiting them to a general overview, the three types of adaptation are:

Physiological Adaptations

These types of adaptations are related to changes in the metabolism of different organisms. Some organs in an animal body function differently when certain changes occur in the environment. The two most well-known physiological adaptations are hibernation and estivation. These are two different types of inactivity where the metabolic rate slows down so much that the animal can survive without eating or drinking anything.

In both cases, temperature is usually a factor. While we often think of bears hibernating for the winter, it is not only when the temperature is low that these processes occur. When the temperature is below 0 °C or is above 40 °C and there is a relatively low humidity, certain animals can lower their basal metabolic rate for serious amounts of time.

Structural Adaptations

The external structures of animals can also help them to better adapt to their environment. Some of the most obvious examples to us might be an aquatic animal developing fins to swim or a mammal growing thicker fur to survive freezing climates. Some of the most attractive adaptations in nature occur for reasons of crypsis (e.g. camouflage) and mimicry.

Cryptic animals are those which camouflage perfectly with their environment and are almost impossible to detect. Certain reptiles and insects such as chameleons and stick insects may spring to mind as being particularly good at this type of animal adaptation. Another reason for this and many other adaptations is for evasion of predators. For example, the monarch butterfly is poisonous which leads to not having many predators. The viceroy has a very similar physical appearance, but is not poisonous. It is not often predated upon either.

Behavioral Adaptations

These adaptations lead the animals to develop certain behavioral characteristics to survive the individual or the species. Fleeing from a predator, hiding during sleep, seeking refuge from climate issues or moving to find different food sources are all behavioral adaptations. The two most characteristic forms of behavioral adaptations are probably migration and courtship. Migration allows the animals to find better resources or evade threat. Courtship is a set of behavioral patterns with the desired result of finding a mate and reproducing.

Examples of Adaptations in Animals

Here are some further examples of how animals have made themselves adaptable to the specific environment in which they live.

Land Adaptation

Shells of certain animal embryos such as reptiles and birds are examples of adaptation to a terrestrial environment. The hard shell prevents the embryo from dying by providing physical protection outside of the body. Fur in mammals is another important adaptation as it allows them to protect their vital organs and maintain function in cold climates.

Aquatic Adaptation

Fins and flippers on various types of fish and aquatic mammals have evolved as adaptations to better survive in water. Likewise, the interdigital membranes of amphibians and birds (the webbing between their 'toes') have the same adaptive purpose.

Light Based Adaptation

Many nocturnal animals have developed eyeballs which allow them to see at night. The lack of light means they need this ability to avoid predators and find their own food sources. However, some animals have gone the other way. The golden mole has eyes, but they are completely covered as they spend all of their time underground and no longer have any use for them.

Temperature Adaptation

The accumulation and maintenance of fat under the skin is a type of adaptation in cold climates According to Allen's rule, animals that live in cold areas have shorter extremities, ears, tails and snouts than animals that live in warmer areas. This is so then can preserve heat. In warmer climates, animals such as the fennec fox have larger ears for the opposite reason. These large appendages are used to cool down the body.

Limits to Animal Aadaptation

Animal adaptations are various and extensive. They allow species to survive, but eventually something will come along to replace them. This is due to the limits of their adaptive abilities. However, many adaptations are not being allowed to happen naturally as humans are exerting such an influence in the world that they bypass natural selection.

A report in 2012 from the Royal Society shows that numbers of a species are key in their success of adaptation as does a range of genetic variation. Since humans have killed so many animals to near extinction, many species struggle to return their species to the numbers they once enjoyed. Also, the hazardous effects of climate change are making it difficult for animals to adapt in time to the rapidly changing environment.

Animal Learning

Animal learning is the alternation of behaviour as a result of individual experience. When an organism can perceive and change its behaviour, it is said to learn.

That animals can learn seems to go without saying. The cat that runs to its food dish when it hears the sound of the cupboard opening; the rat that solves a maze in the laboratory; the bird that acquires the song of its species—these and many other common examples demonstrate that animals can learn. Yet what is meant by saying that animals can learn? What, in other words, is learning? This question proves exceedingly difficult to answer, and, in fact, some theorists propose that no single, all-encompassing definition of learning is at all possible. Moreover, a moment's reflection yields the realization that there exist different kinds of learning. The learning of number concepts, for example, surely seems to be of a different nature than the learning of the association between the sound of a cupboard door and the receipt of food.

General Nature of Learning

Many animals live out their lives following fixed and apparently unvarying routines. Among numerous species of solitary insects, for example, the life cycle consists of the following unvarying events: the females lay their eggs on a particular plant or captured prey; the newly hatched larvae immediately start eating and then follow a standard sequence of developmental stages; the adults recognize appropriate mates by a set of fixed signs, perform a fixed sequence of mating responses, provision their eggs with suitable nourishment, and finally die before the next generation hatches. The same unchanging sequence is repeated generation after generation. And it is, of course, eminently successful. The same set of responses is invariably elicited by the same set of stimuli, because those responses were, and continue to be, adaptive. Where circumstances do not change, there is little need for an animal's behaviour to change. Even many aspects of the behaviour of mammals show a similar fixity. A dog withdraws its foot if it is pricked and a young child his hand if burned; both people and rabbits blink whenever an object is moved rapidly toward their eyes; the feeding behaviour of young infants of virtually all mammalian species consists of sucking elicited by contact with the lips.

Whenever the same response is always appropriate in a particular circumstance, there is little reason why an animal should need to learn what to do in that circumstance. But the world is not always so stable a place. The food supply that was plentiful yesterday may be exhausted today, and the foraging animal that always returns to the same spot will starve to death. Moreover, a particular food supply may be temporarily depleted but will be replenished if left long enough; the successful forager needs to remember where the supply was and when it was last visited, so as to time a return to advantage. In other words, circumstances may change, and the same response is not always appropriate to the same stimuli. Knowing what behaviour is appropriate may depend, therefore, on keeping track of past events.

Possible Explanations of Behavioral Changes

If an animal's behaviour toward a particular stimulus changes, one must look for an explanation of that change. One possible explanation is that the change is due to learning, but there are numerous other possibilities. If a definition of learning is to be provided, that definition must specify when to attribute the change to learning, and when to other causes.

At least two other major causes of behavioral change have been widely recognized. The first of these is motivation. A laboratory rat may pick up, chew, and swallow a pellet of food at one moment; half an hour later, after having eaten 20 grams of food, the rat will simply ignore any further pellets offered. Similarly, a male rat may mount and copulate with a receptive female introduced into his cage, but he will not repeat this pattern of behaviour endlessly even if offered the opportunity to do so. Some male territorial birds, such as chaffinches, will feed amicably beside other males at certain times of day or certain seasons of the year, but at other times they will launch an attack on any intruding male. In all these cases, it is more reasonable to attribute the change in behaviour not to anything the animal has learned but rather to a change in the creature's motivational state.

It should not be thought, however, that just because all of these examples can be attributed to a single item—i.e., motivation—that their detailed explanation will always be the same. The analysis of motivation is itself a large field of study, and it has proved to be more profitable to concentrate on the specific explanation of individual cases of changes in behaviour rather than to search for broad explanatory principles that end up being nearly vacuous. Nonetheless, it does seem possible to draw a contrast between motivational explanations for such changes and those that appeal to learning.

A second broad class of changes in behaviour can be attributed to maturation. We are inclined to ascribe the unfolding pattern of behaviour that emerges over the first few weeks of life to this ill-defined process. Newborn rat pups, for example, are relatively helpless; their eyes do not open for about two weeks, and their main sources of sensory input are probably touch and smell. As their sensory apparatus matures, the pups exhibit changed behavioral responses. The other obvious instance of a maturational change in behaviour is that which comes with sexual maturity: sexually mature adults of most species behave toward one another in ways quite different from those of younger members of the species. It is not only courtship and mating behaviour that change with sexual maturity; for instance, male puppies urinate in the same way as females, by squatting, and it is the onset of sexual maturity that produces the adult pattern of cocking a hind leg.

The concept of maturation is probably no better defined than that of motivation, and it is equally important to stress that it must cover a number of different processes. And, as with motivation, it is more profitable to analyze each case in detail, in order to uncover the precise mechanisms involved, than it is simply to label a change as an example of maturation. Indeed, at the level of physiological process, it seems probable that both motivational and maturational changes are often due to alterations in the hormonal state of the animal, and the distinction between the two is largely one between the unidirectional nature of the change in the case of maturation contrasted with the cyclical change common to short-term motivational states.

But how are these changes discerned from those that might be ascribed to learning? In many cases, of course, the answer is because a precise causal explanation has been provided: a great deal is known, at a physiological level, about the changes in brain and body associated with the motivational states of hunger and thirst. Even without any such detailed knowledge of the underlying mechanisms, it is possible to insist that certain changes in behaviour be attributed to motivation rather than to learning if the opportunity to learn anything relevant was lacking and the opportunity for a motivational change was present. If, for example, an animal that has been deprived of food for a long time behaves in one way toward food-related stimuli, but some hours later, after having been given ample opportunity to eat, it behaves differently toward those stimuli, the obvious interpretation is a motivational one. This interpretation would be strengthened if the animal had not come into contact with these stimuli during the intervening period and had been given, as far as one could judge, no other opportunity to learn anything about them. Learning, in other words, depends on certain kinds of opportunity, and a definition of learning may well turn out to be no more than a specification of the particular set of opportunities and experiences that produce it.

Circumstances that Produce Learning

A particular change in behaviour is attributed to learning, then, because it is possible to specify the set of circumstances that produced it. What are those circumstances? It is common to claim that learning depends on practice. (An older generation of experimental psychologists would have claimed that it depended on "reinforced" practice.) This definition can be misleading, however, if it causes one to attribute to learning all behavioral changes that follow what appears to be practice. In other words, it is not enough to show that an animal appeared to engage in practice and its behaviour subsequently changed. A temporal correlation of this sort does not establish a causal connection. Young birds, for example, are unable to fly, and their first attempts at flight are clumsy and ill-coordinated. Casual observation suggests that young birds improve with practice, gradually perfecting the set of skills they display as adults, but experimental analysis suggests that this practice may be unnecessary. Young birds have been brought up under restricted conditions that completely prevented their flying. When released at the age at which normally reared birds fly proficiently, the experimental subjects flew—without practice—as successfully as those that had spent their time in trial flight. The development of the skill appears to depend more on the maturation of strength and agility than on specific practice.

The notion that learning depends on practice also seems unduly restrictive and is, perhaps, an unnecessary legacy of an earlier version of behaviourism. It is not obvious that an animal should actually have to engage in a particular form of behaviour in order that this pattern of behaviour should be affected by learning. In many cases, indeed, no such practice seems necessary. The young of many songbirds must, it is quite clear, learn their species-typical song. There are several aspects to this learning process, one of which may indeed involve practicing the song at the beginning of the young bird's second season. But another critical aspect is simply exposure to the adult song at some point during the autumn of the young bird's first year, at a time when the young bird does not practice singing at all. Deprived of such experience, chaffinches and song sparrows produce an extremely impoverished version of the adult song; some finches may develop a song more characteristic of another species if that is what they heard during this period of their life. There are numerous other examples where learning appears to depend more on the opportunity to observe than on the opportunity for practice.

This suggests that the definition of learning will have to refer to changes in behaviour that are attributable to particular kinds of experience. The danger now is that, as with motivation and maturation, the definition of learning will be so broad and vague as to be useless. As in those cases, it may be more profitable to concentrate on more detailed analysis of particular instances of learning. Such analysis has, for example, led to widespread agreement on the definition of classical conditioning, a particular type of learning whose study was pioneered by the Russian physiologist Ivan Petrovich Pavlov. In a typical experiment on classical conditioning, an experimenter might arrange a correlation between the ringing of a bell and the delivery of food to an animal. The animal predictably learns to direct food-related activity toward the sound of the bell. Analyses of such experiments have led to the definition of classical conditioning as a type of learning that occurs when there is a correlation between two stimuli and the animal's behaviour toward one of these stimuli changes in a predictable manner determined by the nature of the other. This definition, is useful because it specifies both the circumstances responsible for learning (a temporal correlation between two stimuli) and the general way in which experience of those circumstances changes behaviour (the animal starts directing toward one stimulus responses that are related to those normally directed toward the other). Experimental psychologists and ethologists, however, have devised a tremendous range of procedures for studying learning in animals. The range and variety are such that it may be well-nigh impossible to formulate a meaningful definition of the circumstances that produce learning, for the definition either will be so restrictive that it clearly applies to only a fraction of the cases that should be regarded as instances of learning, or it will be so broad that it says nothing.

Rather than pursue any further the attempt to find an all-embracing, single definition of learning, it seems more useful to provide narrower definitions for particular cases, along the lines suggested above for classical conditioning. One consequence of this approach is that it may encourage the belief that learning consists of a large number of distinct processes that have nothing in common with one another. It is, of course, an open question as to whether this is true: it is certainly possible that, just as with the concept of motivation, the layman's concept of learning encompasses a large number of different cases whose underlying mechanisms are quite distinct. It is important not to prejudge this issue. Insistence on a single, global definition may well tend toward just such prejudgment by encouraging the belief that learning is a single, common process. To start by drawing some distinctions between types of learning does not rule out the possibility of seeing whether the various cases studied do have anything in common.

In the final analysis, as is true of all scientific definitions, the definition of learning is a matter of theory. It has been said that a good scientific definition is the end product of good theory and experiment, not the starting point. Thus, there is a single process of learning if it turns out to be possible to devise a single theory that adequately accounts for the variety of cases in which learning is assumed to occur. Superficial appearances may be deceptive: just because the circumstances that produce learning in two cases, along with the consequences of that learning, appear quite different, it does not follow that the processes underlying learning are different. For instance, the phenomenon of filial imprinting, first seriously analyzed by the Austrian ethologist Konrad Lorenz, appears to be a highly specialized form of learning in which a newborn animal (e.g., a chick, duckling, or gosling) rapidly learns to follow the first salient, moving object it sees. Normally this object will be the mother, but Lorenz discovered that the range of potential imprinting objects is large, extending from Lorenz himself to a bright red ball. There is no question but that some process of learning occurs here, and Lorenz assumed it to be highly specialized. Yet one theory seeks to explain imprinting in terms of simple classical conditioning. Whether or not the account of imprinting provided by this theory is correct, the point is made that how learning is defined and whether it is defined as a single, monolithic process or as many specialized processes are, in the end, questions of theory.

Types of Learning

Simple Nonassociative Learning

When experimental psychologists speak of nonassociative learning, they are referring to those instances in which an animal's behaviour toward a stimulus changes in the absence of any apparent associated stimulus or event (such as a reward or punishment). Studies have identified two major forms of simple nonassociative learning, which are to some extent mirror images of one another: habituation and sensitization.

Habituation

A classic example of habituation is the following observation on the snail Helix albolabris. If the snail is moving along a wooden surface, it will immediately withdraw into its shell if the experimenter taps on the surface. It emerges after a pause, only to withdraw again if the tap is repeated. But continued repetition of the same tapping at regular intervals elicits a briefer and more perfunctory withdrawal response. Eventually, the stimulus, which initially elicited a clear-cut, immediate response, has no detectable effect on the snail's behaviour. Habituation has occurred.

Habituation can be defined in behavioral terms as a decline in responding to a repeatedly presented stimulus. As such, it is a very widespread phenomenon, one that can be observed in animals ranging from single-celled protozoans to humans. Most animals behave differently to novel and familiar stimuli: the former sometimes elicit startle responses, sometimes investigatory or exploratory responses; the latter often apparently are ignored. The suggestion that habituation is a simple form of learning, however, implies that it can be distinguished from some even simpler potential causes of this sort of change in behaviour. One reason why an animal might stop responding to a stimulus is that it no longer detects the stimulus; i.e., some form of sensory adaptation might have occurred. Another potential cause is fatigue: perhaps some temporary refractory state is produced by repeated elicitation of the same response, making it impossible to perform that response again. Whether or not one would want to call either of these processes a form of learning is doubtful. But both behavioral and physiological evidence establishes that habituation cannot be explained in these terms.

The critical behavioral evidence is that habituation can be disrupted by almost any change in the experimental conditions. If repeated presentation of one stimulus leads to habituation of a response, the same response can still be elicited by a different stimulus. Even if the experimenter presents a novel stimulus that does not itself elicit the response in question, its presentation may restore the response on the next trial in which the originally habituated stimulus is presented. This latter observation, usually referred to as an instance of dishabituation, seems to rule out any simple sensory adaptation; both observations rule out simple effector fatigue.

Neurophysiological analysis of habituation in various mollusks—for example, in the sea snail Aplysia—has confirmed that habituation need not depend on changes in the activity of sensory or motor neurons. In the case of Aplysia, researchers have studied the gill withdrawal reflex, a response that rapidly habituates to repeated stimulation of the snail's siphon or mantle shelf. But habituation still occurs even if it is elicited by direct, electrical stimulation of the motor nerve, bypassing the sensory receptors completely; and recording from the sensory nerve during normal habituation reveals no decline in its level of activity. These observations eliminate sensory adaptation as a possible cause of the animal's having ceased to respond to the stimulus. Effector fatigue can be ruled out by showing that direct stimulation of the motor neurons controlling the withdrawal response can still elicit a perfectly normal reaction even after the response has completely habituated. Research shows that habituation in Aplysia depends on changes in the activity of more central neurons. Repeated tactile stimulation of the siphon, leading to habituation of the withdrawal response, causes changes in the activity of the motor neurons innervating the response. Specifically, these motor neurons show a decline in excitatory postsynaptic potential, which is the electrical change that enables the nerve impulse to cross the gap (synaptic cleft) that separates one neuron in the pathway from the next. The decline in excitatory postsynaptic potential short-circuits the response. Moreover, the presentation of a novel stimulus, sufficient to dishabituate the behavioral response, restores the postsynaptic potential.

Habituation occurs even in animals without a central nervous system—probably in single-celled protozoans; certainly in animals such as the coelenterate Hydra, which have a diffuse nerve net and do not appear to be capable of associative learning. Among mammals, habituation of certain reflex responses can be observed even in "spinal" subjects, that is, those whose spinal cord has been severed from the brain. There can be little doubt, then, that habituation is not only widespread, but that it also can be a relatively simple phenomenon. There is, however, no guarantee that it is the same phenomenon wherever it appears. The waning response to a repeatedly presented stimulus admits of a number of different explanations. In principle, as we have already seen, it might be due to sensory adaptation, effector fatigue, or a more central neural change. These distinctions make rather little sense in the case of a single-celled animal. And one should not necessarily expect the habituation observed in a spinal mammal to involve precisely the same mechanisms as those responsible for comparable behavioral effects in an intact animal. Some psychologists have proposed theories of habituation that appeal to processes of classical conditioning. Such a theory is not likely to apply to the habituation observed in an animal that shows no capacity for classical conditioning.

Habituation is usually, as here, classified as an instance of simple, nonassociative learning. It is supposedly nonassociative because all that happens in the course of habituation is that a stimulus is repeatedly presented and the animal's behaviour changes; there is, on the face of it, no other event with which the stimulus can be associated. Habituation must therefore, it appears, be understood by reference to some change in the pathway between stimulus and response, and the work with Aplysia and other mollusks shows how this analysis may proceed at the physiological level. But if habituation is not always the same phenomenon, it is possible that different processes may underlie the habituation of the startle response to a loud noise in an intact mammal. And despite appearances to the contrary, those processes may involve some associative learning. One suggestion is that novel stimuli elicit a biphasic response: an initial increase in startle responses, which include components of emotion or anxiety, followed by a rebound in the opposite direction. Habituation occurs when the latter, rebound response becomes conditioned to the stimulus, occurring sooner and sooner with each repetition of the stimulus and thereby damping down and eventually canceling out the initial reaction. An alternative possibility is that long-term habituation depends on associating the repeatedly presented stimulus with the context in which it occurs, a suggestion that would explain why presentation of the stimulus in a different context sometimes leads to dishabituation.

The generality of habituation implies that this behavioral phenomenon has considerable adaptive significance; if true, it would be quite reasonable to expect that a number of different mechanisms might have evolved to produce the behavioral result. The adaptive value of habituation is not difficult to see. A novel stimulus may signify danger, and an animal should react to this stimulus either by withdrawing or at least by orienting toward it to see what will happen next. But if the same stimulus occurs again with no further consequence, it is probably safe: regular repetition of the same stimulus implies that it is part of the background, such as the waving of a branch in the wind or the shadow caused by a piece of seaweed floating with the waves. If the stimulus is not dangerous, time should not be wasted on it. Withdrawal, especially in the case of a snail into its shell, is a time-consuming effort, incompatible with such vital activities as searching for food. If it is important, therefore, for animals to be wary of novel stimuli, it is equally important that they should discriminate the novel and potentially dangerous from the familiar and probably safe.

Sensitization

The effect of habituation is to eliminate unnecessary responses, but the main function of learning has usually been thought to be the production of new responses. Traditional psychological theories of learning have assumed that the learning of new patterns of behaviour comes about through the association of a new response with a particular stimulus. Consequently, psychologists usually have either ignored the possibility that nonassociative processes might be sufficient to increase the probability of a new response or regarded it as a nuisance that interferes with the measurement of associative changes. They have rarely treated it as a subject worthy of study in its own right.

This is unfortunate, for the nonassociative phenomenon of sensitization is probably fairly widespread, and it provides a simple means of acquiring adaptive behaviour. Sensitization is said to occur when the repeated presentation of a particular significant stimulus (such as food or electric shock) lowers the threshold for the elicitation of appropriate behaviour to the point where a second stimulus, not normally capable of calling forth that behaviour, now does so. A typical example is provided by the behaviour of the marine worm Nereis. If the worm is kept in a small tube and fed at regular intervals, it becomes progressively more likely to respond to any novel stimulus, such as a change in illumination, by exploratory, food-seeking movements toward the open end of the tube. If, on the other hand, the worm receives mild electric shocks at regular intervals, it becomes progressively more likely to respond to a novel stimulus by withdrawal.

The first point to note about sensitization is its relationship to habituation. Habituation refers to a decline in the probability of responding to a repeatedly presented stimulus. Sensitization, by contrast, refers to an increase in the probability that behaviour appropriate to a repeatedly presented stimulus will occur, even in response to another stimulus. Although these two outcomes cannot be observed simultaneously, it is quite possible that the same operation—repeated presentation of a stimulus—can simultaneously engage two different processes, one causing a decline in the probability of responding to that stimulus, the other causing an increase. Experimental analysis suggests that both processes are real and may be engaged in the same experiment, so that the observed change in behaviour actually results from a mixture of the two. Typically, the process of habituation wins out, and what is observed is an overall decline in responding. But a common finding in habituation experiments is that responding initially increases before declining; the implication is that the initial presentations of a stimulus result in more sensitization than habituation, while further presentations produce more habituation than sensitization. A second factor influencing the relative importance of the two processes is the intensity or significance of the stimulus. A weak stimulus, or one with little intrinsic biological significance, will show relatively rapid habituation and little or no initial sensitization. A stronger stimulus, especially one, such as food or shock, that has substantial significance to the animal, may show marked sensitization and relatively little habituation.

The second point about sensitization is that it may mimic the effect of associative learning or conditioning. As has been mentioned, in a classical conditioning experiment a neutral stimulus, such as a change in illumination, is paired with the delivery of a significant stimulus, such as food or shock. Repeated pairing causes the neutral stimulus to elicit responses initially called forth by the significant stimulus; for example, a change in illumination that has been associated with an electric shock would come to elicit retreat or withdrawal. But in the case of the worm Nereis, experiments demonstrate that the light would come to elicit this change in behaviour whether or not it had been paired with shock: all that is needed is sufficient exposure to the shock. To attribute the change in behaviour toward the light to its association with food or shock, one must show that this change is greater than that which would have resulted from sensitization alone.

The physiological processes underlying sensitization, like those underlying habituation, have been analyzed in experiments on such invertebrate species as Aplysia. Not surprisingly, the mechanisms involved appear to mirror one another. Whereas habituation is correlated with a decline in postsynaptic potentials, sensitization is correlated with an increase in the magnitude of postsynaptic potentials at the same locus.

Although sensitization has often been treated as a nuisance whose effects must be controlled in studies of habituation or associative learning, it remains a process worthy of study in its own right, for the behavioral changes it produces can have significant adaptive value. Without requiring the presumably more complex neural machinery necessary to subserve associative learning, sensitization enables animals to respond to local variations in the occurrence of significant events. If an animal's sources of food tend to occur together (that is, they are not distributed randomly in time or space), then it pays that animal, having once found food, to continue to behave in a food-gathering manner. Conversely, the animal that is increasingly wary after exposure to danger will have a better chance of evading a lurking predator. Sensitization thus enables an animal to take advantage of statistical regularities in the occurrence of significant events, without requiring it to detect other events that predict the significant ones. No doubt, further advantage accrues to the animal that can perform such calculations, for associative learning provides a powerful means of predicting the future. But there can be equally little doubt that such a process requires a more elaborate nervous system.

Associative Learning Conditioning

The study of animal learning in the laboratory has long been dominated by experiments on conditioning. This domination has been resisted by critics, who complain that conditioning experiments are narrow, artificial, and trivial, and, as such, miss the point of what animals are adapted to learn. From the critics' point of view, one unfortunate effect of their attacks has been the progressive refinement and elaboration of the theory of conditioning to the point where it can often explain the exceptions to which they drew attention. This is not to insist that associative learning is the sole, or even the most important, form of learning in vertebrates, but rather to introduce the idea that the processes underlying conditioning may be more interesting than older theories and an earlier generation of textbooks suggested.

Classical and Instrumental Conditioning

Pavlov was not the first scientist to study learning in animals, but he was the first to do so in an orderly and systematic way, using a standard series of techniques and a standard terminology to describe his experiments and their results. In the course of his work on the digestive system of the dog, Pavlov had found that salivary secretion was elicited not only by placing food in the dog's mouth but also by the sight and smell of food and even by the sight and sound of the technician who usually provided the food. Anyone who has prepared food for his pet dog will not be surprised by Pavlov's discovery: in a dozen different ways, including excited panting and jumping, as well as profuse salivation, the dog shows that it recognizes the familiar precursors of the daily meal. For Pavlov, at first, these "psychic secretions" merely interfered with the planned study of the digestive system. But he then saw that he had a tool for the objective study of something even more interesting: how animals learn. From about 1898 until 1930, Pavlov occupied himself with the study of this subject.

Pavlov's experiments on conditioning employed a standard, simple procedure. A hungry dog was restrained on a stand and every few minutes was given some dry meat powder, an event signaled by an arbitrary stimulus, such as the ticking of a metronome. The food itself elicited copious salivation, but, after a few trials, the ticking of the metronome, which regularly preceded the delivery of food, also elicited salivation. In Pavlov's terminology, the food is an unconditional stimulus, because it invariably (unconditionally) elicits salivation, which is termed an unconditional response. The ticking of the metronome is a conditional stimulus, because its ability to elicit salivation (now a conditional response when it occurs in reaction to the conditional stimulus alone) is conditional on a particular set of experiences. The elicitation of the conditional response by the conditional stimulus is termed a conditional reflex, the occurrence of which is reinforced by the presentation of the unconditional stimulus (food). In the absence of food, repeated presentation of the conditional stimulus alone will result in the gradual disappearance, or extinction, of its conditional response. In translation from the Russian, the terms "conditional" and "unconditional" became "conditioned" and "unconditioned," and the verb "to condition" was soon introduced to describe the experimental activity.

To the American psychologist Edward L. Thorndike must go the credit for initiating the study of instrumental conditioning. Thorndike began his studies as a young research student, at about the time that Pavlov—already 50 years old and with an eminent body of research behind him—was starting his work on classical conditioning. Thorndike's typical experiment involved placing a cat inside a "puzzle box," an apparatus from which the animal could escape and obtain food only by pressing a panel, opening a catch, or pulling on a loop of string. Thorndike measured the speed with which the cat gained its release from the box on successive trials. He observed that on early trials the animal would behave aimlessly or even frantically, stumbling on the correct response purely by chance; with repeated trials, however, the cat eventually would execute this response efficiently within a few seconds of being placed in the box.

Thorndike's procedures were greatly refined by another U.S. psychologist, B.F. Skinner. Skinner delivered food to the animal inside the box via some automatic delivery device and could thus record the probability or rate at which the animal performed the designated response over long periods of time without having to handle the animal. He also adopted some of Pavlov's terminology, referring to his procedure as instrumental, or operant, conditioning; to the food reward as a reinforcer of conditioning; and to the decline in responding when the reward was no longer available as extinction. In Skinner's original experiments, a laboratory rat had to press a small lever protruding from one wall of the box in order to obtain a pellet of food. Subsequently, the "Skinner box" was adapted for use with pigeons, who were required to peck at a small, illuminated disk on one wall of the box in order to obtain some grain.

In experiments on both classical conditioning and instrumental conditioning, the experimenter arranges a temporal relation between two events. In Pavlov's experiment the food was always preceded by the conditional stimulus; in Skinner's original experiment the delivery of food was always preceded by the rat's pressing the lever. Conditioning, or associative learning, is inferred if the animal's behaviour changes in certain ways and if that change can be attributed to the temporal relationship between these events. If the dog started salivating to the ticking of a metronome just because it had recently received food, rather than because the delivery of food had been signaled by the metronome, this should be regarded as an instance of sensitization rather than associative learning. One of the simplest ways of establishing that the change in behaviour results from the temporal relationship between the conditional stimulus and the unconditional stimulus in a classical experiment, or between the response and the reinforcer in an instrumental case, is to impose a delay between the two. A gap of even a few seconds between the rat's pressing the lever and the delivery of food will seriously interfere with the animal's ability to learn the connection. And although in some classical experiments evidence of conditioning can be found in spite of relatively long gaps between the conditional stimulus and the unconditional stimulus, increasing this interval beyond a certain point invariably causes a decline in conditioning.

Laws of Associative Learning

The temporal relation between the conditional stimulus and the unconditional stimulus, or between the response and the reinforcer, was for a long time regarded as the primary determinant of conditioning. Conditioning is certainly a matter of associating temporally related events, but temporal contiguity is only one of several factors—and probably not the most important—that influences conditioning. A variety of experiments have shown that classical conditioning will occur only if the conditioned stimulus is the best predictor of the occurrence of the unconditional stimulus. In other words, it is the correlation between two events, just as much as their temporal contiguity, that establishes an association between them. A pigeon, for example, will learn by classical conditioning to peck an illuminated disk in a Skinner box if, whenever the disk is illuminated, food is delivered. This temporal relationship between the light and food can be preserved intact, but if the experimenter now arranges that food is equally available at other times (when the light is not on), the pigeon will not peck at the illuminated disk. Delivering food at other times destroys the correlation between light and food (although leaving the temporal relationship untouched) and abolishes conditioning.

Although some conditioning will occur when the conditional stimulus is not perfectly correlated with the delivery of food (perhaps because on a proportion of trials the conditional stimulus is presented alone without food) or when the temporal relationship is less than perfect (there is a gap between the conditional stimulus and the delivery of food), this conditioning is abolished if the experimenter ensures that there is some better predictor always available. If a dog is conditioned to the ticking of a metronome paired with the delivery of food, the animal will salivate in response to the metronome even if the food is presented in no more than 50 percent of the trials. If, however, a light is illuminated on those trials when the metronome is accompanied by food, and not on the remaining 50 percent of the trials, the dog will become conditioned to the light and not to the metronome. Similarly, a pigeon will learn to peck at a disk illuminated with red light even if a gap of several seconds separates this response from the delivery of food. But if, during this interval, after the red light has been turned off and before food is delivered, a green light is turned on, the pigeon will never learn to peck at the red light. It is as though the pigeon attributes the occurrence of food to the most recent potential cause (now the green light rather than the red), and the dog attributes food to the stimulus best correlated with its delivery (the light rather than the metronome). Conditioning, in other words, occurs selectively to better predictors of reinforcement at the expense of worse predictors. This same principle explains the earlier observation of the role of correlation in general. The pigeon will not associate the illumination of the disk with food if food is equally probable both when the light is on and when it is switched off; from the pigeon's point of view, food occurs whenever the animal is placed in the Skinner box. The illumination of the light signals no increase in the probability of food, and the best predictor of food is the mere fact of being in the Skinner box.

Temporal contiguity, therefore, is not necessarily the most important factor in successful conditioning. Moreover, there is yet another factor that should be stressed. It will hardly have escaped the reader's attention that there is an astonishing artificiality to the typical conditioning experiment conducted by Pavlov or Skinner. An animal is placed in a bare, confined space; lights are flashed on and off; the animal is permitted to operate some mechanical contrivance; some meat powder or a pellet of food is delivered. How could one possibly suppose that the ways in which animals learn anything of importance in the real world will be illuminated by this contrived and restrictive kind of experiment? This question raises large issues. But one point should be acknowledged right away: the more restricted the range of experimental manipulations employed, the greater the chance that the investigator will completely miss important principles. Experiments with lights and metronomes failed to reveal the following important principle of conditioning: animals appear to have built-in biases toward associating some classes of stimuli with certain classes of consequences. The most dramatic instance of this principle is provided by conditioned food aversions. If rats eat some novel-flavoured substance and shortly thereafter are made mildly ill (for example, by an injection of a drug such as apomorphine or lithium chloride), they afterward will show a marked aversion to the novel food. Because they will show an aversion even though an interval of several minutes, or sometimes even hours, intervenes between eating the food and the onset of the illness, there has been some question as to whether this should be regarded as an instance of conditioning at all. But the parallels between food aversions and other forms of conditioning are so extensive that it is hard to believe that some common processes are not involved. And there is no question but that the length of the interval is important; other things being equal, rats will form a stronger aversion to a food they have eaten recently than to one they have eaten several hours earlier.

The most interesting feature of such aversions is that they are, by and large, confined to foods. If rats suffer the unpleasant experience of being made ill, they are not likely to show an aversion to anything other than a novel-tasting food or drink they have recently ingested. As in other forms of conditioning, the novelty of the potential conditional stimulus is important. Rats will not show any marked aversion to a thoroughly familiar diet unless the experience of illness is repeatedly induced shortly after eating the daily ration, just as, in Pavlov's experiments, conditioning will proceed only slowly to the ticking of a metronome if the dog has heard this sound repeatedly before. The more striking restriction, however, is that it is the taste of the food or drink that is associated with illness. If rats drink plain tap water before being made ill, they will show little aversion to tap water (since there is no novelty here). But even if a novel buzzer is sounded while they are drinking and they are then made ill, they will not associate the buzzer with the illness. This is certainly not because rats are unable to associate the buzzer with an aversive consequence. If drinking water while the buzzer is sounded produces a mild electric shock, rats will rapidly learn to stop drinking whenever they hear the buzzer. In this case it is the flavour of the water that rats find difficult to associate with the shock; punishing rats with a mild shock whenever they drink sugar-flavoured water has little effect on their tendency to drink sugar-flavoured water. The flavour of food or drink is readily associated with subsequent illness, but only poorly associated with other painful consequences. Conversely, an external stimulus such as a buzzer or flashing light, which is readily established as a signal for shock, is only with great difficulty associated with illness.

The full explanation of this finding remains uncertain. It is known that even very young rats show such selectivity, so it cannot depend solely on any prior experience. What is easy to see is that this behaviour makes biological sense. Internal malaise, such as that caused in the psychologist's experiment by an injection of lithium, will in the real world usually be a consequence of eating spoiled or poisonous food or of drinking tainted water. The most reliable sign of such food or drink will be its taste, and animals predisposed to associate the taste of what they have ingested with subsequent illness are likely to be better equipped to avoid potentially harmful food in the future. On the other hand, painful injury, mimicked in the laboratory by a brief electric shock, is hardly likely to be a consequence of eating food of a particular flavour; it will usually be caused by external circumstances, such as contact with a sharp or very hot object or a narrow escape from a predator. The natural suggestion is that the function of conditioning is to enable animals to find out what causes certain events of biological significance. If this is so, a built-in bias toward associating certain classes of events together makes adaptive sense. Conditioning is not just a matter of associating two events because one happens to follow the other; it is more profitably seen as the process whereby animals discover the most probable causes of events of consequence to themselves.

Laws of Performance

Conditioning could have no function at all, however, if it did not involve changes in an animal's behaviour. Nor could scientists infer that conditioning has occurred unless they could observe, at some point, a change in an animal's behaviour attributable to certain conjunctions of events. So, although conditioning may involve the formation of associations between events or the attribution of particular events to their most probable antecedent causes, it must also include some mechanisms for translating these associations into changes in behaviour.

For an earlier generation of behaviourists, the fundamental fact about conditioning was precisely that it changed behaviour, and the theories they advanced were determined by this fact. The description of conditioning as the establishment of a new response to a stimulus that had not previously elicited that response naturally suggested that conditioning was a matter of forming new stimulus–response connections. This conceptualization led to the development of the stimulus–response theory, variations of which long provided the dominant account of conditioning. One version of the stimulus–response theory suggested that the mere occurrence of a new response to a given stimulus, as when Pavlov's dog started salivating shortly after the metronome had started ticking, is in itself sufficient to strengthen the connection between the two. Thorndike, however, argued that the probability that a particular stimulus will repeatedly elicit a particular response depends on the perceived consequences of this response. According to this view, new stimulus–response connections are strengthened only if the response is followed by certain kinds of consequences.

There are several questions raised here, and it is important to keep them distinct. One is whether responses are sometimes (or even always) modified by their consequences. Although denied by some theorists, their denial seems distinctly paradoxical. A rat whose presses on a lever are followed by the delivery of a food pellet will press the lever again; if the only consequence of pressing the lever is the delivery of a painful shock, the rat will desist from this action. Thorndike's law of effect—which stated that a behaviour followed by a satisfactory result was most likely to become an established response to a particular stimulus—was intended to summarize these observations, and it is surely an inescapable feature of understanding how and why humans and other animals behave. In keeping with this understanding, parents reward children for good behaviour and punish them for bad. When this fails to produce the desired behaviour, we are inclined to argue that the child is finding other sources of reward or does not find the intended punishment particularly unpleasant, or that the parents' behaviour is hopelessly inconsistent. We are far less likely to question the assumption that, other things being equal, people (and other animals) repeat actions that have desirable consequences and avoid repeating those that have undesirable consequences.

Thorndike's law of effect was, however, also a theory of how reward and punishment modify behaviour. This theory, which states that behaviour normally is modified by changing the strength of stimulus–response connections, finds less general acceptance today. A simple experiment suggests one reason for this. A rat is trained to press a lever in a Skinner box, being rewarded with a small quantity of sucrose solution for each press of the lever. Once the response has been established, the rat is removed from the Skinner box. The next day, while in its home cage, the animal is given sucrose solution to drink and shortly thereafter is made ill by an injection of lithium. Once this treatment has established a strong aversion to the sucrose, the rat is returned to the Skinner box, where, despite the opportunity to do so, the animal does not press the lever again. The result is hardly surprising: there is no reason to expect the rat to perform a response whose sole consequence is the delivery of the now aversive sucrose solution. But this behaviour cannot be explained by Thorndike's theory, for according to Thorndike all that the rat learned in the first stage of the experiment was a new stimulus–response habit; stimuli from the Skinner box should, by Thorndike's reasoning, now elicit the response of pressing the lever. Thorndike's stimulus–response theory credits the rat with no acquired knowledge of the connection between pressing the lever and obtaining sucrose; the function of sucrose is merely to strengthen the stimulus–response connection.

That responses are modified by their consequences, therefore, need not call for Thorndike's theoretical account of this fact. It is probably more reasonable to suppose that animals learn about the relationship between their actions and consequences (just as they can also learn about the relationship between any other classes of events), and that they then modify their actions in accordance with the current value of these consequences. The next question to consider is whether this is an entirely general principle of performance, or whether it applies only to some classes of response in some kinds of situations. Why, for example, does Pavlov's dog start salivating to the ticking of the metronome? Is it because the response of salivating is followed by a rewarding consequence? The response is, at first, elicited by the sight of food and is shortly followed by the rewarding consequence of chewing and swallowing the food. But another simple experiment suggests that salivating to the metronome is not strengthened because it is followed by food. The experimenter can turn on the metronome for five seconds on each trial, at the end of which time the dog receives food—but only if it did not start salivating before the arrival of food. Now the response of salivating to the metronome is followed by an undesirable consequence, the cancellation of the food that would otherwise have been delivered on that trial, but the dog still cannot help salivating (at least sometimes) to the metronome. The implication is that salivating is not a response modified by its consequences, but one reflexly elicited by food and also by any stimulus associated with food. Voluntary responses can be modified by their consequences; involuntary responses (such as blushing when a person is embarrassed or the release of adrenalin when a person is angry or afraid) cannot.

The reason Pavlov's dog starts salivating to the metronome is, just as Pavlov himself supposed, that the association between metronome and food means that the metronome can substitute for food. To put it another way, the metronome now produces activity in neural centres normally responsive to the delivery of food, activity that is reflexly connected to the salivary response.

It should not be thought that only autonomic, glandular responses are involuntary in this sense. If a small light is always illuminated for five seconds before the delivery of food to a hungry pigeon, the pigeon will learn, by classical conditioning, to approach and peck at the light. Exactly the same experiment as that described above can be undertaken, with food delivered only on those trials when the pigeon does not approach and peck the light during the initial five seconds. The pigeon cannot help doing so. Pavlovian conditioning appears to be a widespread phenomenon, applying to a relatively wide range of responses.

Functions of Conditioning

The behaviour of the dog and pigeon in the above experiments seems maladaptive, precisely because it violates the law of effect. If the way to obtain food is to refrain from performing a particular response, then that is what the law of effect says the animal should do. The law of effect makes obvious adaptive sense; several writers, indeed, have pointed to the analogy between the law of effect and natural selection. Just as natural selection favours those variations that happen to increase fitness, so the law of effect selects those responses that happen to be followed by certain consequences.

The fact that Pavlovian conditioning may result in apparently maladaptive behaviour in the artificial confines of the experimental psychologist's laboratory, however, does not mean that it is not adaptive in the real world. The pigeon's behaviour provides a clue. In a normal classical conditioning experiment, where the illumination of a small light regularly precedes the delivery of food, the pigeon will rapidly learn to approach and direct pecks at the light. Approach and pecking are food-related activities: what is happening is that a simple process of Pavlovian conditioning is ensuring that responses related to food are being elicited by stimuli associated with food. It is not difficult to appreciate the adaptive significance of a process that results in animals approaching places where they have found food in the past, or in learning that a particular novel object is in fact an example of food, and directing food-related activity toward these stimuli in the future.

Pavlovian conditioning also affects other significant behaviours. For example, it probably provides the basic process by which animals learn to avoid poisonous foods. If a novel food is associated with illness, its taste will elicit responses of disgust or nausea, ensuring that the substance will subsequently be rejected after the first taste. In territorial birds and fish, aggressive displays and attacks can become conditioned to stimuli that regularly precede the appearance of a rival male. A male already primed to threaten and attack an intruder, because he has learned that certain signs herald the appearance of the intruder, should be more successful in defense of his territory than the male that is unprepared. Experimental analysis has, in fact, nicely confirmed this expectation. In general, any pattern of defensive behaviour that is adaptive in response to an intruder or predator—such as displaying or fighting, fleeing or taking other evasive action, or freezing into immobility or feigning death—will be even more adaptive if performed in advance, at the first reliable signal of the predator's or intruder's appearance.

The process of Pavlovian conditioning thus often enables animals to behave appropriately in anticipation of events of biological significance, without involving any direct modification of that behaviour by its success or failure. But further modification must sometimes be of further advantage. For instance, it is not always enough just to approach a stimulus associated with food; if that stimulus is a prey species, it may take evasive action that will require much more elaborate behaviour on the part of the predator. This can be seen in the feeding behaviour of the oystercatchers, a group of birds that eat bivalve mollusks. Oystercatchers first catch their pray by probing down the hole made by the bivalve in the mud; the sight of the hole must be rapidly established as a conditional stimulus for food. But the birds must then perform a complex series of actions to get at the mollusk's flesh, and this skilled sequence of responses also must be learned, presumably in accordance with the law of effect. Similarly, many animals have a wide range of defensive behaviour patterns; in the laboratory, at least, which one eventually predominates in any given situation normally depends on which one successfully enables the animal to escape or to avoid aversive consequences. In all these cases, it appears that instrumental conditioning serves to modify, via the law of effect, initial responses that owed their origin to Pavlovian conditioning.

The adaptive value of instrumental conditioning is an area of research that has seen some fruitful collaboration among experimental psychologists, ethologists, and behavioral ecologists. From ecology has come the "optimal foraging theory," the idea that efficient foraging behaviour should maximize an animal's net rate of food intake. From ethology and experimental psychology has come the idea that an animal's instrumental behaviour in any given situation is a product of competition between various possible activities, a competition whose resolution depends on weighing the costs and benefits of increasing one activity at the expense of another. Both in the laboratory and in more natural settings, for example, the proportion of time spent searching for one kind of food depends not only on the probability of finding that food and on its value when found but also on the probability of the animal finding an alternative food if it looks elsewhere. There is also abundant evidence that animals improve their foraging efficiency with practice; this clearly must depend on learning which stimuli signal the availability of which kinds of food, the most efficient way of taking a given food, and the most effective distribution of time between alternatives.

Spatial Learning

One of the major problems many animals must confront is how to find their way around their world—for example, to know where a particular resource is and how to get to it from their present location, or what is a safe route home to avoid a predator. Such spatial learning may cover only the highly restricted confines of an animal's home range or territory, or it may embrace a migration route of several hundreds or even thousands of miles. Although some forms of navigational behaviour may be explicable in relatively simple terms, not necessarily requiring appeal to processes more complex than those of simple conditioning, others suggest some quite new principles.

Maze Learning

In the psychologist's laboratory, the primary method of studying spatial learning has been to put a rat in a maze and watch how it finds its way to the goal box, where it is fed. As befits the analytic (some would say sterile) approach so popular in experimental psychology, the elaborate and complex mazes used in earlier studies (the very first published experiment used a scaled-down replica of the maze at Hampton Court, London) soon gave way to something very much simpler, a T-maze or Y-maze. A rat placed at the end of one arm must run to the central choice-point, from where it has to enter one of the two remaining arms. Although extremely simple, even this apparatus allows for a number of possible modes of solution. One possibility is that the rat learns to execute a particular response, a left turn or a right turn, at the choice-point, because that response is followed by food. A second possible solution is that the rat learns that the two alternative arms differ in some particular way and further learns to associate one of the arms with food and hence to choose it. The third and most interesting possibility is that the rat learns to define the rewarded arm not in terms of its own intrinsic characteristics but by its spatial relationship to an array of landmarks outside the maze. Thus the rat might learn that the correct arm is the one pointing to the left of a window and away from a table with a lamp on it. Experiments show that whenever such landmarks are available, this third solution mode is the one used.

Perhaps the most convincing demonstration that rats can find their way to a particular location—one defined solely in terms of its spatial relation to various external landmarks—has been provided by experiments in which the animals are placed in a large circular tank of water and must swim to a transparent platform submerged somewhere in the middle of the tank. They can rapidly learn to do this, regardless of where they are initially put into the tank and even though the platform itself is invisible. (The invisibility of the platform is shown by the following: if the platform is moved, the rat will swim straight past it, heading instead toward the position it used to occupy.)

Rats in these experiments are not simply approaching a single landmark; they locate their goal by reference to its spatial relationship with a whole series of landmarks, no one of which is necessary. This can be established by using half a dozen arbitrary but easily identified objects as landmarks during maze training. Removal of any one or two of them in no way disrupts the rat's behaviour.

If all the landmarks are systematically rotated around the room, the rat will identify a new arm of the maze as correct (the one that has the same relationship to the landmarks as the initially correct arm). If, however, the landmarks are rearranged in such a way as to destroy their original spatial relationship to one another, the rat does not know which arm to choose.

The processes involved in this sort of learning are not well understood. Some psychologists have been sufficiently impressed by the rat's flexibility in these experiments to argue that the animal is constructing a map of its environment—not, obviously, a written map but an internal, maplike representation that encodes a complete set of spatial relationships between major landmarks. The best evidence for such a maplike representation would be if a rat could take an unfamiliar route when its original route to a goal is blocked. Unfortunately, there is little evidence of such performance in rats, except in the not especially critical case where the goal, or a stimulus very close to it, is clearly visible from the choice-point. On the other hand, studies of long-range navigation have shown that some animals can do just this.

Navigation

Salmon return from the ocean to spawn in the stream in which they were hatched; swallows return to the same nest sites in northern Europe each spring from wintering in southern Africa. These and other examples of large-scale migrations have long fascinated students of animal behaviour, and experimental intervention has produced some remarkable results. A Manx shearwater was taken in an airplane from its breeding site on the island of Skokholm, off south Wales, to Boston, Mass. It returned to Skokholm within 13 days of being released in Boston; the direct distance between these two points is 3,050 miles, which implies (assuming that the bird did not fly at night) a minimum average speed in excess of 20 miles per hour. An albatross flew from a release site in the Philippines to its home in Midway Island, a direct distance of 4,120 miles, in 32 days.

How do these animals navigate across such great distances? Numerous cues have been implicated in different instances. Near to home, animals probably rely on local cues quite different from those used at a distance. For example, experiments show that salmon distinguish their home streams on the basis of smell, although this sense can hardly come into play while the fish are swimming in the open ocean. Other investigations have demonstrated that diurnal birds use visual information derived from the position of the Sun, while those that migrate at night rely on the pattern of the stars. There have been several suggestions that certain long-range migrators are sensitive to the Earth's magnetic forces; sensitivity to auditory cues has also been suggested in some cases.

The most intensive analysis of long-range navigation has been undertaken with homing pigeons. These birds are trained by being released from sites progressively further from their home loft. Just what the pigeons learn on these training flights is not entirely clear. In part, they obviously learn the visual landmarks immediately surrounding the home loft, but experimental evidence suggests that they use such landmarks only very close to home. Once some training has been given, however, a pigeon can be taken 100 miles or more in any direction from home, and it will, within a few minutes of its release, start flying in a homeward direction.

One general class of theory on homing behaviour postulates that the pigeon detects a discrepancy between a particular set of stimuli observed at the release site and its stored knowledge of what that set of stimuli should be like at home, and it then flies in such a direction as to reduce this discrepancy. Different versions of this theory appeal to different sets of stimuli that might be used to guide the pigeon home. At one time, a popular idea was that the pigeon used the Sun's height in the sky in combination with an internal clock. At any given season and time of day, the Sun's height in the sky—and, by extrapolation from its current rate of climb, its maximum height—are unique to a single place (in this case, the pigeon's home). Assuming that the pigeon's home loft and the release site are both in the Northern Hemisphere, then if the Sun's maximum height is lower at the release site, the release site is north of home; if higher, then the release site is south of home. If the Sun will reach its maximum height later than at home, the release site is west of home; if earlier, the site is east of home. If released at noon at a site in the Northern Hemisphere 200 miles northeast of home, the pigeon must fly so as to raise the maximum height of the Sun (i.e., south), and so as to stop the Sun falling (i.e., west).

This explanation is immensely ingenious and, although calling for some astonishingly fine sensory discriminations on the part of the pigeon, not impossible in principle. Unfortunately, it is probably wrong. Two critical experiments have produced results quite at variance with its predictions. The first suggested that pigeons do not rely on the height of the Sun to navigate at all. In this experiment, the pigeons were confined to a laboratory from which they could see the Sun for only a relatively short time around noon each day, and the apparent height of the Sun above the horizon was raised or lowered by allowing the birds to view the Sun only through a complex series of mirrors. This should have had drastic effects on their perception of the true position of their home; for instance, an increase of 70′ in the apparent height of the Sun at noon would correspond to an 80-mile southward relocation of the home. The pigeons were then taken from home and released 40 miles south, where they saw the real Sun for the first time in several weeks. If the Sun's height was indeed a critical stimulus in navigation, the pigeons would be expected to fly south rather than north. In fact, they correctly flew north.

The second experiment involved shifting the birds' internal clock, by confining them indoors and exposing them to a new light–dark cycle. Independent observations had shown that this procedure is entirely successful: if a bird is confined indoors for a few weeks with the lights switched on every day at midnight and switched off at noon, its clock soon will be entrained on this new cycle, so that 6:00 AM is regarded as the middle of the day. In the critical experiment, the bird was taken out of the laboratory and released at 6:00 AM (true time) from a site 50 miles south of home. The Sun, now seen for the first time in several weeks, was just rising; but, according to the pigeon's internal clock, the time at home was noon. This implied that the release site was a long way west of home, and if the pigeon were using the height of the Sun as a cue to guide it home, it should have flown east. In fact, the pigeon flew west.

The result of the second experiment indicates that the pigeon was using the position of the Sun in the sky, and that the clock shift had been effective (for the pigeon was not flying in the direction of home). This is readily explained by the hypothesis that the pigeon used the Sun as a compass. If we allow, for the sake of argument, that the pigeon knew that the release site was south of home, then it should have tried to fly north. In the Northern Hemisphere at noon, the Sun is due south; therefore, the pigeon—whose internal clock said it was noon—should fly away from the Sun. But although the pigeon's shifted clock said that it was noon, the true local time was 6:00 AM, and the Sun was in the east. Flying away from the Sun, the pigeon flew west. This experiment then suggests first, that the pigeon was not using the height of the Sun at all; second, that it used the Sun's horizontal position, or azimuth, to provide a compass bearing; and, third, and most important, that the pigeon had some other map that told it that the release site was south of home. In general, a compass is of no use without a map.

The basis for the map component of the pigeon's navigational skill remains extremely obscure. There is evidence from studies of many migratory birds that the compass component is in some sense innate, but that a map of the relative positions of the summer and winter habitats and of other places in between (or even not in between) develops only with the experience of migration. For example, starlings that breed around the Baltic Sea fly southwest in autumn to winter in southern England, northern France, and Belgium. When captured during this autumn migration and released in Switzerland (some 500 miles south of their normal route), experienced, adult birds flew back to northern France and Belgium—even though they had presumably never flown over any part of this route before. Young birds, however, for whom this was the first migration, flew southwest from Switzerland and ended up in southern France or northern Spain. They clearly had a compass that told them which direction was southwest; what they lacked was any knowledge of the spatial relationship between their present location in Switzerland and their goal in northern France.

Perceptual Learning

According to Thorndike's stimulus–response theory, learning, which is reducible to the strengthening and weakening of the tendency to perform a particular response in the presence of a particular stimulus, occurs only when that response is performed; learning, in other words, depends on trial and error. Even in the realm of simple conditioning, there are good reasons to question this restriction. Conditioning is better conceptualized as the acquisition of knowledge about temporal relationships between events rather than as the acquisition of behaviour. Spatial learning seems to be a matter of learning about spatial relationships between objects and places in one's environment and, apparently, the construction of some sort of map that will subsequently permit the animal to perform a new sequence of actions across unknown territory.

Imitation and Observational Learning

One reason why Thorndike adopted such a narrow, behavioral view of learning was that he looked for evidence of other forms of learning without success. Having taught one cat to escape from the puzzle box by operating a latch, he looked to see whether a second cat would acquire the correct solution simply by watching the first. A series of such experiments produced uniformly negative results, and Thorndike concluded that trial and error was the only form of learning available to animals other than humans.

Why Thorndike should have been so unsuccessful is something of a mystery, for later experiments have established quite convincingly that animals can often benefit from watching another member of their species perform a particular task. Casual observation in natural settings, for instance, reveals that young chimpanzees intently watch their elders perform intricate tasks; this certainly suggests that learning by observation is very common in some species.

Experimental analysis has revealed a number of important distinctions concerning the role of observation in behaviour. For example, domestic chickens that have eaten to satiation a particular source of food will start eating again if they observe other chickens feeding. Although the observation of conspecifics engaged in a particular activity has clearly affected the tendency of the satiated chicken to engage in that activity, it is not clear what they might have learned from this observation. They already know how to peck, and they already know that the grain before them is palatable food. It is probably more appropriate to regard this as an instance of "social facilitation" and to say that one of the stimuli that elicits feeding in chickens is the sight of other chickens feeding.

The example above demonstrates the minimum requirement for establishing that an animal has learned by observation: In the absence of the opportunity to observe another, the animal must have been unlikely to have performed a particular response, and the reason for this must reside in lack of knowledge. An artificial, laboratory example of observational learning would be to allow an observer rat to watch a demonstrator rat pressing a lever for food. If the observer has never before pressed a lever and, given the opportunity, now does so much more rapidly than another rat denied the opportunity to observe the demonstrator, surely some genuine observational learning has occurred. But even here it remains difficult to establish exactly what it is that the observer has learned by watching the demonstrator, and more elaborate experiments may be required to elucidate this. An experiment with two monkeys showed how this may be done. The monkeys took turns acting as demonstrator and observer. The demonstrator's task was to choose between two objects, one of which contained some hidden food. Since the objects were changed on each new trial for the demonstrator, there was no way for the animal to know which choice was correct, and it necessarily picked one at random. The observer, however, could watch the demonstrator's trial and thus could find out which of the two objects in a particular set was correct. Given an opportunity to choose between the two, the observer more often than not chose correctly. That the observer was not simply watching the demonstrator, but was in fact looking to see the outcome of the choice, is established by the finding that the observer performed somewhat more accurately on those trials when the demonstrator's choice was wrong than on those when it was right.

This last finding points to a further distinction, that between observing the actions of another and imitating those actions. In this particular experiment, the monkeys clearly were not imitating one another, or they would have copied each other's choices even when these were wrong. A demonstration of imitation is provided by the behaviour of oystercatchers feeding on mussels. Having found a mussel, an adult oystercatcher obtains the food from within either by inserting its beak in the right place and cutting the muscle that holds the shell together or by pecking a hole in the weakest point of the shell. Young birds develop the method employed by their parents, but experiments in which chicks were fostered by adults with a different habit from that of the natural parents have established that this behaviour is not genetically determined. Rather, the young birds imitate the actions they observe being performed by their foster parents.

The best known natural example of such imitation was provided by a troop of macaques in Japan. In order to lure the monkeys out of the forest and into the open, where their behaviour could be better studied, scientists routinely left sweet potatoes and wheat on the beach. The monkeys ate this food but clearly disliked the fact that it had become liberally mixed with sand. A young female member of the troop, however, discovered that sweet potatoes could readily be washed free of sand, and that a handful of wheat and sand could be thrown into a pool, where the sand would sink, leaving the wheat floating behind. Both customs spread through the troop, first to the immediate family and young companions of the original inventor, and last of all (an interesting touch) to the old, conservative males. Other examples of observational learning are readily apparent in the behaviour of animals in the field, but in many cases, as in some of the laboratory studies cited above, it remains difficult to elucidate just what it is that has been learned.

Song Learning

A special case of observational learning is that of young birds acquiring their species-typical song. Numerous species of animals, including many birds, produce species-typical calls or other vocalizations as adults; in many cases, however, there is little evidence that learning plays any significant role in their development. In many species of crickets, for example, the song is stereotyped, and the pattern of neural activity that produces the song can be detected even in young animals who neither sing nor apparently react to the adult song. But in most songbirds, there is reason to believe that learning has a significant effect on the development of the adult song.

The interesting feature of this learning is that it sometimes occurs in two distinct phases separated by several months. The first of these can be regarded as purely observational learning, the second as the perfection of the song through practice (i.e., as imitation of a model). Song sparrows, for example, do not develop a normal adult song unless they have the opportunity to hear the song during their first autumn. There is thus a sensitive period during which they must hear their species' song if they are to develop normally, but it is important to note that they do not themselves sing at all during the first autumn. It is not until the next spring that they start practicing the song. At this point, they do not need to hear other sparrows singing, but they do need to hear themselves. If the bird is deafened before it starts practicing, only a very crude song emerges. The implication is that, during exposure in the first autumn, the sparrow learns to identify the detailed song and establishes a template of it; the following spring, the sparrow starts singing and needs practice to match its output to the stored template.

The song sparrow provides an example of a particularly clear separation between observation and imitation. In other species, such as the chaffinch, the young bird learns from exposure to song in the first autumn, but refinement of the song is produced by further exposure to other chaffinches singing during the following spring. In yet others, such as indigo buntings, the adult bird learns its song from territorial neighbours. But even where there is no temporal separation between the two aspects of learning, it still seems valid to distinguish between the learning involved in establishing the template and that involved in perfecting the motor skill.

If song learning consists solely of the young bird learning to reproduce the adult, species-typical song, one might wonder why any learning should be necessary at all. Why should the song not develop simply through maturation, or, in other words, why is not the template, at least, genetically laid down in the bird's brain? In fact, studies indicate that a relatively crude template is innately determined in most species. There are very strict limits to the range of songs that a bird of one species can learn. Moreover, among chaffinches and certain other species, even if a young bird hears no song at all it will still develop a crude song that has recognizable features of the full, species-typical one. The degree of this innate specification varies widely from species to species: at one extreme are such birds as cuckoos, which develop a standard call with no prior exposure at all; at the other extreme are such birds as marsh warblers, which develop idiosyncratic songs picked up, it seems, from any other species they come in contact with during the sensitive period.

Species whose song acquisition involves a great deal of individual learning are probably those in which individual birds develop slightly different songs. In some species, such as song sparrows, there are recognizable local "dialects" that the young birds learn from adults living in the same region. In other species, there is even more variation between individuals. If one function of the song is to attract a mate, then an interplay is called for between a song that simply advertises the singer's species and one that establishes his individual identity. The importance of individual learning, then, depends on the role of the song in the mating patterns of the species.

Imprinting

The young of many species are born relatively helpless: in songbirds, rats, cats, dogs, and primates, the hatchling or newborn infant is wholly dependent on its parents. These are altricial species. In other species, such as domestic fowl, ducks, geese, ungulates, and guinea pigs, the hatchling or newborn is at a more advanced stage of development. These are precocial species, and their young are capable, among other things, of walking independently within a few minutes or hours of birth, and therefore of wandering away from their parents. Since mammals are dependent on their mothers for nourishment, and even birds are still dependent on parental guidance and protection, it is important that the precocial infant not get lost in this way. The phenomenon of filial imprinting ensures that, in normal circumstances, the precocial infant forms an attachment to its mother and never moves too far away.

Although imprinting was first studied by the Englishman Douglas Spalding in the 19th century, Konrad Lorenz is usually, and rightly, credited with having been the first not only to experiment on the phenomenon but also to study its wider implications. Lorenz found that a young duckling or gosling learns to follow the first conspicuous, moving object it sees within the first few days after hatching. In natural circumstances, this object would be the mother bird; but Lorenz discovered that he himself could serve as an adequate substitute, and that a young bird is apparently equally ready to follow a model of another species or a bright red ball. Lorenz also found that such imprinting affected not only the following response of the infant but also many aspects of the young bird's later behaviour, including its sexual preferences as an adult.

Imprinting, like song learning, involves a sensitive period during which the young animal must be exposed to a model, and the learning that occurs at this time may not affect behaviour until some later date. In other words, one can distinguish between a process of perceptual or observational learning, when the young animal is learning to identify the defining characteristics of the other animal or object to which it is exposed, and the way in which this observational learning later affects behaviour. In the case of song learning, observation establishes a template that the bird then learns to match. In the case of imprinting, observation establishes, in Lorenz' phrase, a model of a companion, to which the animal subsequently directs a variety of patterns of social behaviour.

With imprinting, as with song acquisition, one can ask why learning should be necessary at all. Would it not be safer to ensure that the young chick or lamb innately recognized its mother? There are, in fact, genetic constraints on the range of stimuli to which most precocial animals will imprint. A model of a Burmese jungle fowl (the species whose domestication produced domestic chickens) serves as a more effective imprinting object for a young chick than does a red ball; there is even evidence that imprinting in the latter case involves different neural circuits from those involved in imprinting to more natural stimuli. Nonetheless, it is clear that the innate constraints are not very tight and that a great deal of learning normally occurs. The most plausible explanation, as in the case of song learning, is that imprinting involves some measure of individual identification. Lorenz argued that one of the unique characteristics of imprinting was that it involved learning the characteristics of an entire species. It is true that imprinting results in the animal directing its social and mating behaviour toward other members of its own species, and not necessarily toward the particular individuals to which it was exposed when imprinting occurred. But learning usually involves some generalization to other instances, and there does not seem to be anything peculiar to imprinting here. The primary function of imprinting, however, is to enable the young animal to recognize its own mother from among the other adults of its species. This no doubt is particularly important in the case of such animals as sheep, which live in large flocks. Only learning could produce this result.

There is also an important element of individual recognition in at least some cases of imprinting's effects on sexual behaviour. Experiments with Japanese quail have shown that their sexual preferences as adults are influenced by the precise individuals to whom they are exposed at an earlier age. Their preferred mate is one like, but not too like, the individuals on whom they imprinted. The preference for some similarity presumably ensures that they attempt to mate with members of their own species. The preference for some difference is almost certainly a mechanism for reducing inbreeding, since young birds will normally imprint on their own immediate relatives.

The difference between imprinting and song learning lies in the consequences of observational learning. The effect of imprinting is the formation of various forms of social attachment. But what mechanism causes the young chick or duckling to follow its mother? Lorenz thought that imprinting was unrewarded, yet the tendency of a young bird to follow an object on which it has been imprinted in the laboratory can be enhanced by rewarding the bird with food. Rewards also occur outside the laboratory: the mother hen not only scratches up food for her young chicks, she also provides a source of warmth and comfort. Moreover, following is also rewarded by a reduction in anxiety. As chicks develop over the first few days of life, they show increasing fear of unfamiliar objects; they allay this anxiety by avoiding novel objects and approaching a familiar one. This latter object must be one to which they have already been exposed—in other words, one on which they have imprinted. Imprinting works because newly hatched birds do not show any fear of unfamiliar objects, perhaps because something can be unfamiliar only by contrast with something else that is familiar. On the contrary, the newly hatched birds are attracted toward salient objects, particularly ones that move. Once, however, a particular object has been established as familiar and its features identified, different objects will be discriminated from it. These will be perceived as relatively unfamiliar, and hence they will provoke anxiety and the attempt to get as close as possible to the more familiar object. The imprinting of the young bird on one object necessarily closes down the possibility of its imprinting on others, as these will always be relatively less familiar. Thus, there is normally a relatively restricted period in the first few hours or days of life during which imprinting can occur. The only way to prolong this period is to confine the newly hatched bird to a dark box where it is exposed to no stimuli; prevented from imprinting during this period of confinement, the bird imprints on the first salient object it sees after emerging.

Complex Problem Solving

Experimental psychologists who study conditioning are the intellectual heirs of the traditional associationist philosophers. Both believe that the complexity of the human or animal mind is more apparent than real—that complex ideas are built from simple ideas by associating simple elements into apparently more complex wholes. According to this perspective, the only relationship between these ideas is their association, and the determinants of these associations are themselves relatively simple and few in number. Neither conditioning theorists nor associationist philosophers, however, have lacked for critics, who claim that intelligent problem solving cannot be reduced to mere association. Although allowing that the behaviour of invertebrates, and perhaps that of birds and fish, may be understood in terms of instincts and simple forms of nonassociative and associative learning, these critics maintain that the human mind is an altogether more subtle affair, and that the behaviour of animals more closely related to man—notably apes and monkeys, and perhaps other mammals as well—will share more features in common with human behaviour than with that of earthworms, insects, and mollusks.

The idea that animals might differ in intelligence, with those more closely related to humans sharing more of their intellectual abilities, is commonly traced back to Charles Darwin. This is because the acceptance of Darwin's theory of evolution was at the expense of the ideas of the French philosopher René Descartes, who held that there is a rigid distinction between man, who has a soul and can think and speak rationally, and all other animals, who are mere automatons. The Cartesian view had, in fact, been challenged long before Darwin's time by those who believed (as seems obvious from even the most casual observations) that some animals are notably more complicated than others, in ways that probably include differences in behaviour and intelligence. It was, however, the publication of Darwin's Descent of Man that stimulated scientific interest in the question of mental continuity between man and other animals. Darwin's young colleague, George Romanes, compiled a systematic collection of stories and anecdotes about the behaviour of animals, upon which he built an elaborate theory of the evolution of intelligence. It was largely in reaction to this anecdotal tradition, with its uncritical acceptance of tales of astounding feats by pet cats and dogs, that Thorndike undertook his studies of learning under relatively well-controlled laboratory conditions. Thorndike's own conclusions, already noted above, were distinctly Cartesian: animals ranging from chickens to monkeys all learned in essentially the same way, by trial and error or simple instrumental conditioning. Unlike man, none could reason.

This controversy actually involves two questions, which are worth keeping apart. The first is whether theories of learning based on the results of, say, simple conditioning experiments are sufficient to explain all forms of learning and problem solving in animals. The second question is whether new and more complex processes operate only in some animals, that is to say, whether some animals are more intelligent than others. The distinction between these questions is not always easy to preserve, for they are clearly related, and an answer to one usually has implications for the other.

Discrimination of Relational and Abstract Stimuli

Laboratory studies of habituation and conditioning usually employ very simple stimuli, such as lights, buzzers, and ticking metronomes in Pavlov's experiments. Some of the other examples of learning considered earlier have already suggested that animals can actually respond to additional, more complex stimuli. Even the solution of simple spatial discriminations in the laboratory requires the animal to learn about spatial relationships between different landmarks; migration or navigation over hundreds of miles demands abilities at least as complex as this. Song learning requires the young bird to discriminate between different sequences of subtly varying notes and calls, and the individual recognition involved in imprinting requires response to elaborate configurations of features.

Thus, one way in which a problem may become more difficult is if its solution depends on response to more subtle changes in stimuli. Numerous laboratory studies have examined the abilities of a variety of animals to perform such discriminations. The phenomenon of transposition, first studied in chicks by the Gestalt psychologist Wolfgang Köhler, suggests that animals may solve even simple discriminations in ways more complex than the experimenter had imagined. Köhler trained his chicks to perform simple discriminations—say, to choose a large white circle (five centimetres in diameter) in preference to a small white circle (three centimetres in diameter). He then sought to discover whether the animal was responding to the relationship between the two stimuli or to the absolute characteristics of the stimuli. In other words, had the chick learned to select the larger of the two circles, or had it learned to pick the five-centimetre circle? If the former were the case, Köhler reasoned that given the choice between the five-centimetre circle and an even larger one (eight centimetres in diameter), the animal should transpose the relationship and choose the larger circle. This was indeed the result, demonstrating that the animal was responding in terms of the relationship between stimuli rather than, or at least in addition to, their absolute properties.

Transposition experiments show that animals can respond to relationships between stimuli varying along a particular continuum of physical characteristics: size, brightness, hue, etc. Another question is whether animals can respond to an abstract property of a stimulus array, independent of the actual physical stimuli making up that array. In experiments on counting, the animal must choose between an array containing, say, five stimuli and one containing three. The actual stimuli in the array vary from trial to trial, in order to rule out the possibility that the animal is responding in terms of other features, such as differences in total area or brightness, between the arrays. Counting experiments have been tried on birds more frequently than on any other class of animal, and several species, notably ravens, rooks, and jackdaws, have solved this type of problem. This success may not be entirely by chance, for there is reason to believe that the stimulus that controls when a female bird stops laying eggs is something to do with the number of eggs already laid and in the nest. Chimpanzees, however, have been trained to label pictures of various objects (e.g., spoons, shoes, padlocks, and balls) with the numeral specifying the number of objects in the picture. Moreover, rats and other standard laboratory animals have solved similarly abstract discriminations, for example, of temporal duration. A rat can learn to perform one response after a stimulus has been turned on for two seconds and a different response after the stimulus has been turned on for five seconds. The nature of the actual stimuli employed can vary without disrupting the rat's discrimination, suggesting that it is the duration of the stimuli to which the rat responds.

Concept learning makes up another class of discriminations that may be solved by the abstraction of a particular property or set of properties from a very wide array of individual stimuli. In a typical experiment, a pigeon is shown a large number of colour photographs of natural scenes: half of these contain, somewhere within the scene, all or part of a tree or group of trees; the other half contain no tree (although there might be flowers, a climbing rose, or other plants). Responding to the pictures of trees is rewarded, but responding to the remaining pictures is not. Pigeons rapidly learn the discrimination. In one sense, perhaps this is not surprising: birds that roost in trees, one is inclined to argue, must be able to recognize them. But pigeons can learn other discriminations with almost equal facility; for example, they can be trained to distinguish between underwater scenes containing a fish and similar views with no fish present. In such cases, the class of stimuli in question is one for which their evolutionary history can hardly have prepared pigeons. The question, of course, is how the pigeons solve such problems. Are they, in some sense, abstracting a conceptual rule for categorizing the world into classes of stimuli? Or are they responding to what is no doubt a very large number of particular features that differentiate trees or fish from other objects in the world?

Pigeons, in common with most birds, rely more heavily on vision, and certainly have better developed colour vision, than most mammals—with the exception of primates. There is evidence that monkeys can solve the concept discriminations that have been set to pigeons, but there is no evidence that other mammals can. For extensive comparative analysis, therefore, it is necessary to turn to different kinds of tasks. One that has been studied almost to excess is discrimination reversal. In reversal tasks, an animal is first trained on a simple discriminative problem: for example, to choose the left-hand arm of a T-maze, where it is rewarded, rather than the right arm, where it is not. Once the animal has solved the problem, the experimenter reverses the reward assignments, so that the food is now in the right arm rather than the left. Training continues until the animal has learned this reversal, whereupon the assignment of reward is switched back to the left arm. And so on. Rats trained on this series of reversals eventually become extremely adept at the task. Although the initial reversal causes considerable problems, with animals making many more errors than on the original discrimination, after a few more reversals these difficulties vanish. Eventually, rats solve each new reversal in fewer trials than they took to solve the original discrimination, often with no more than a single error.

Similarly efficient performance has been observed in a relatively wide range of mammals. More interesting was the early suggestion that the few species of fish (goldfish, African mouthbreeders, and Paradise fish) trained on similar problems showed no evidence of the increase in efficiency displayed by mammals. The fish would learn the first reversal slowly and laboriously, and the 20th reversal equally slowly. Subsequent experiments have established that this was an unfairly pessimistic assessment, for improvements in experimental techniques have been accompanied by a significant improvement in the fish's performance, a finding that highlights the extreme difficulty of assessing the relative efficiency of widely differing animals on supposedly the same task. Nevertheless, it remains doubtful that goldfish are as adept at reversal tasks as rats are.

The theoretical question, however, is how rats attain such efficiency. What processes allow them eventually to learn the reversal of a discrimination faster than they originally learned the discrimination itself, and often with only a single error? The most plausible suggestion is that they develop a "win–stay, lose–shift" strategy. They learn, in other words, to characterize the alternatives between which they must choose not in terms of their physical features but in terms of whether or not they chose it on the previous trial. They then learn that, if the alternative they chose on the last trial was rewarded, choice of that alternative will be rewarded again on the current trial; while, if it was not, choice of the other alternative will now be rewarded. A variety of other experiments have shown that rats can rapidly learn to use the outcome of one trial to predict the outcome of the next, and hence keep track of regular sequential dependencies in the availability of food or other rewards.

Generalized Rule Learning

Second only to the reversal task in popularity as a tool for the comparative analysis of learning has been the learning set task. The latter is designed to measure the animal's ability "to learn to learn"—in other words, to discover whether after having learned a new behaviour the animal can then more readily learn other related behaviours. For example, an animal is trained on a simple discrimination between two objects, A and B. Once the problem has been solved, the experimenter substitutes a new pair of objects, C and D, for the original pair; when the animal has solved this new problem, yet another new pair, E and F, is substituted, and so on. Rhesus monkeys trained on such a series of problems become progressively more efficient at solving each new problem. Like rats trained on reversal tasks, the monkeys eventually solve each new problem after a single trial, choosing at random on the first trial with each new pair of stimuli but thereafter selecting with essentially perfect accuracy.

Performance on learning sets, as on reversals, was once thought to discriminate between more intelligent and less intelligent animals. Apes and rhesus monkeys were extremely efficient at such tasks, more so even than New World monkeys, who were, in turn, more efficient than any nonprimate mammals. Again, however, there are grave difficulties in the way of making valid comparisons. Primates have better developed visual systems than most other mammals, so it is not surprising that they should be better at solving a series of visual discrimination problems. Even the difference in performance between rhesus and cebus monkeys (Old World versus New World monkeys) turns out to be attributable to differences in colour vision more than anything else. Rats appear to solve learning set tasks very efficiently if olfactory stimuli are used.

Nevertheless, there may be important intellectual differences also underlying the differences in performance. One reason for thinking so arises from consideration of the processes probably involved in mastering learning sets. The win–stay, lose–shift strategy that explains the progressive improvement in reversal learning can also explain the same improvement in the learning set task—but only if the animal can generalize the strategy to novel stimuli. Successful performance requires that the animal learn that the alternative chosen on the last trial, and the outcome of that choice, predict which alternative will be rewarded on this trial, whatever the nature of the alternatives. Some evidence suggests that primates can generalize rules of this sort more readily than many other animals can. Monkeys trained on a series of reversals of a single discrimination will learn the reversal of any new discrimination with equal facility. By contrast, cats trained on comparable problems show little evidence of such transfer.

A discriminative problem widely used in the study of transfer is the "matching-to-sample" discrimination. A pigeon, for example, is required to choose between two disks, one illuminated with red light and the other with green light. The correct alternative on any one trial depends on the value of a sample stimulus, which is also part of each trial. If this third light is red, then the red disk is correct; if green, then green is correct. The correct alternative is the one that matches the sample. Although naturally more difficult than the simple red–green discrimination, matching-to-sample discriminations are learned readily enough by a wide variety of animals; however, there appear to be differences among animals in their capabilities to transfer this learning to a new set of stimuli. Primates and dolphins have shown good evidence of such transfer, but pigeons have shown at best only limited transfer. If pigeons are trained with two or three colours to the point where they are responding with essentially no errors, a substitution of a new colour for one of the trained colours may result in a complete breakdown in the discrimination; there is even some question as to whether they can learn a new matching-to-sample discrimination with new stimuli any faster than pigeons with no prior experience of matching problems.

The abilities to respond in terms of certain relationships between stimuli, to abstract those relationships and invariant features from a complex and changing array of stimuli, and, above all perhaps, to transfer such learning to a completely novel set of physical stimuli seem to be some of the more important processes underlying the solution of complex discriminative problems. The fact that certain evidence suggests that animals may differ in some of these abilities has implications for studies of other forms of problem solving.

Insight and Reasoning

Köhler's best known contribution to animal psychology arose from his studies of problem solving in a group of captive chimpanzees. Like other Gestalt psychologists, Köhler was strongly opposed to associationist interpretations of psychological phenomena, and he argued that Thorndike's analysis of problem solving in terms of associations between stimuli and responses was wholly inadequate. The task he set his chimpanzees was usually one of obtaining a banana that was hanging from the ceiling of their cage or lying out of reach outside the cage. After much fruitless endeavour, the chimpanzees would apparently give up and sit quietly in a corner, but some minutes later they might jump up and solve the problem in an apparently novel manner—for example, by using a bamboo pole to rake in the banana from outside or, if one pole was not long enough, by fitting one pole into another to form a longer rake. Other chimpanzees reached the banana hanging from the ceiling by using a wooden box, or a series of boxes stacked precariously on top of one another, as a makeshift ladder.

Köhler believed that his chimpanzees had shown insight into the nature of the problem and the means necessary to solve it. According to Köhler's interpretation, the solution depended on a perceptual reorganization of the chimpanzee's world—seeing a pole as a rake, or a series of boxes as a ladder—rather than on forming any new associations. But subsequent experimental analysis has cast some doubts on Köhler's claims. The critical observation is that the sorts of solutions that Köhler took as evidence of insight quite clearly depend on relevant prior experience. Chimpanzees will not fit two poles together to form a rake or stack boxes up to form a ladder unless they have had a great deal of prior experience with those objects. This experience may well occur during play, when the young chimpanzee discovers that using a stick can extend the reach of an arm, or that standing on a box can put one within reach of high objects. Thus, what Köhler was studying, without knowing it, was probably the transfer of earlier instrumental conditioning to new situations. As we have already seen, the ability to transfer an old solution to a new stimulus situation is an important one, relevant to a wide range of problem-solving activities. This ability is not at all well understood, but it will not necessarily be greatly illuminated by describing it as insight. Certainly it is not a process unique to the great apes: if the component tasks are sufficiently well-structured, even pigeons can put together two independently learned patterns of behaviour to solve a novel problem.

Combining information from separate sources to reach a new conclusion is one form of reasoning. The paradigm case of reasoning is the solution of syllogisms; for example, when we conclude that Socrates is mortal given the two separate premises that Socrates is a man and that all men are mortal. Employing transitive inference, we can use the premises that Adam is taller than Bertram and that Bertram is taller than Charles to conclude that Adam must be taller than Charles. Reasoning has often been regarded as a uniquely human faculty, one of the few factors, along with the possession of language that distinguishes us from the rest of the animal kingdom.

But are humans the only animals that can reason? The unsatisfying answer must be that it depends on what is meant by reasoning. In a very general sense, most animals appear perfectly able to arrive at a conclusion based on combining information obtained on two separate occasions. A formal demonstration is provided by an experiment on instrumental conditioning. If rats learn that pressing a lever provides sucrose pellets and later learn that eating sucrose pellets makes them ill, they will subsequently put these two pieces of information together and refrain from pressing the lever. Monkeys and chimpanzees, however, have been trained to solve problems that appear more similar to transitive inference. They are first given discriminative training between pairs of coloured boxes, called, for example, A, B, C, D, E. Confronted with the choice between A and B, they learn that choice of A is rewarded and B is not. When B and C are the alternatives, they learn that B is correct; when C and D are the alternatives, C is correct; and so on. Although choice of A is always rewarded, and that of E never is, the remaining three boxes each are associated equally often with reward and with nonreward. Nonetheless, given a choice between B and D on a test trial, the animals choose B.

Syllogistic and transitive inferences are not the only forms of reasoning: humans also reason inductively or by analogy. Indeed, analogical reasoning problems (black is to white as night is to —?) form a staple ingredient of some IQ tests. One chimpanzee, a mature female called Sarah, was tested by David Premack and his colleagues on a series of analogical reasoning tasks. Sarah previously had been extensively trained in solving matching-to-sample discriminations, to the point where she could use two plastic tokens, one meaning same, which she would place between any two objects that were the same, and another meaning different, which she would place between two different objects. For her analogical reasoning tasks, Sarah was shown four objects grouped into two pairs, with each pair symmetrically placed on either side of an empty space. If the relationship between the paired objects on the left was the same as the relationship between those on the right, her task was to place the same token in the space between the two pairs. Thus in one series of geometrical analogies, a simple problem would display a blue circle and a red circle on the left and a blue triangle and a red triangle on the right; the correct answer, of course, was same. But Sarah was equally correct on more complex problems, even when the relationships in question were functional rather than simply perceptual. For example, she correctly answered same when the two objects on the left were a tin can and a can opener and the two on the right a padlock and a key.

Solution of analogies requires one to see that the relationship between one pair of items (whether they are words, diagrams, pictures, or objects) is the same as the relationship between a different pair of items. If simple matching-to-sample requires animals to see that one comparison stimulus is the same as the sample and another is different, solving analogies requires them to match relationships between stimuli. The difficulties encountered in training pigeons to generalize simple matching-to-sample discriminations does not encourage one to believe that they would find analogies very easy.

Language Learning

The ability to speak was regarded by Descartes as the single most important distinction between humans and other animals, and many modern linguists, most notably Noam Chomsky, have agreed that language is a uniquely human characteristic. Once again, of course, there are problems of definition. Animals of many species undoubtedly communicate with one another. Honeybees communicate the direction and distance of a new source of nectar; a male songbird informs rival males of the location of his territory's boundaries and lets females know of the presence of a territory-owning potential mate; vervet monkeys give different calls to signal to other members of the troop the presence of a snake, a leopard, or a bird of prey. None of these naturally occurring examples of communication, however, contains all of the most salient features of human language. In human language, the relationship between a word and its referent is a purely arbitrary and conventional one, which must be learned by anyone wishing to speak that language; many words, of course, have no obvious referent at all. Moreover, language can be used flexibly and innovatively to talk about situations that have never yet arisen in the speaker's experience—or indeed, about situations that never could arise. Finally, the same words in a different order may mean something quite different, and the rules of syntax that dictate this change of meaning are general ones applying to an indefinite number of other sequences of words in the language.

During the first half of the 20th century, several psychologists bravely attempted to teach human language to chimpanzees. They were uniformly unsuccessful, and it is now known that the structure of the ape's vocal tract differs in critical ways from that of a human, thus dooming these attempts to failure. Since then, however, several groups of investigators have employed the idea of teaching a nonvocal language to apes. Some have used a gestural sign language widely used by the deaf to communicate with one another; others have used plastic tokens that stand for words; still others have taught chimpanzees to press symbols on a keyboard. All have had significant success, and several apes have acquired what appears to be a vocabulary of several dozen, and in some cases 100 or 200, "words."

Washoe, a female chimpanzee trained by Beatrice and Allan Gardner, learned to use well over 150 signs. Some apparently were used as nouns, standing for people and objects in her daily life, such as the names of her trainers, various kinds of food and drink, clothes, dolls, etc. Others she used as requests, such as please, hurry, and more; and yet others as verbs, such as come, go, tickle, and so on. Sarah, the chimpanzee trained by Premack to use plastic tokens as words, also apparently learned to use tokens for nouns, verbs (give, take, put), adjectives (red, round, large), and prepositions (in, under). But do these signs or tokens really function as words? Does the ape using them, or obeying instructions from a trainer who uses them, really understand their meaning? Or is the ape simply performing various arbitrary instrumental responses in the presence of particular stimuli because she had previously been rewarded for doing so?

There can be little doubt that chimpanzees do have some understanding of what their "words" refer to. Sarah responded appropriately with her token for red if asked the question "What colour of apple?" both when an actual red apple was shown as part of the question and when only the token for an apple (which happened to be a blue triangle) was presented. To Sarah, the blue triangle surely stood for, or was associated with, the red apple. In another study, after two chimpanzees had been taught the meaning of a number of symbols for different kinds of food and different tools, they were able not only to fetch the appropriate but absent object when requested to do so, but they could also sort the symbols into two groups, one for foods and one for tools. In another series of studies, a pygmy chimpanzee named Kanzi demonstrated remarkable linguistic abilities. Unlike other apes, he learned to communicate using keyboard symbols without undergoing long training sessions involving food rewards. Even more impressive, he demonstrated an understanding of spoken English words under rigorous testing conditions in which gestural clues from his trainers were eliminated.

Kanzi's Primal Language describes researchers' efforts to teach language to a pygmy chimpanzee named Kanzi.

Human language is more than a large number of unrelated words: in accordance with certain implicitly understood syntactic rules, humans combine words to form sentences that communicate a more or less complex meaning to a listener. Can apes understand or use sentences? Undoubtedly they can put together several gestures or tokens in a row. A chimpanzee named Lana, who was trained to press symbols on a keyboard, could type out "Please machine give Lana drink"; Washoe and other chimpanzees trained in gestural sign language frequently produced strings of gestures such as "You me go out," "Roger tickle Washoe," and so on. Skeptical critics, however, have raised doubts about the significance of these strings of signs and symbols. They have pointed out, for example, that when Lana pressed a series of coloured symbols on her keyboard, it was humans who interpreted her actions as the production of a sentence meaning "Please machine give Lana drink." Might it not be equally reasonable to say that she learned to perform an arbitrary sequence of responses in order to obtain a drink? Pigeons can be trained to press four coloured keys—red, white, yellow, and green—in a particular order to obtain food. Psychologists do not feel any temptation to interpret this behaviour as the production of a sentence. What is it about Lana's behaviour that requires this richer interpretation?

In the case of apes trained to use sign language, two other doubts have been raised. First, there is some reason to believe that a disappointingly high proportion of the apes' gestures may be direct imitations of gestures recently executed by their trainers. Second, a sequence of gestures interpreted as a single sentence is often just as readily interpreted as a number of independent gestures, each prompted, in turn, by a gesture from the trainer. Both these conclusions are based on careful examinations of video recordings of interactions between trainers and apes. Whether they will turn out to be generally true remains an open, and heatedly debated, question.

Without any explicit training, apes have nevertheless learned to produce strings of two or three signs in certain preferred orders: "more drink" or "give me," for example, rather than "drink more" or "me give." Do the animals understand that a string of signs in one order means something different from the same signs in a different order? The following anecdote is suggestive. A chimpanzee called Lucy was accustomed to instructing her trainer, Roger Fouts, by gesturing "Roger tickle

Lucy." One day, instead of complying with this request, Fouts signed back "No, Lucy tickle Roger." Although at first nonplussed, after several similar exchanges Lucy eventually did as asked. A simple instance of this sort proves little or nothing, but it may suggest what is needed—namely, that Lucy should understand that changing the order of a set of signs alters their meaning in certain predictable ways. She must generalize the rule that the relationship between the meanings of the signs A-B-C and C-B-A (the same signs in reverse order) is similar to the relationship between the meanings of certain other triplets of signs in her vocabulary when their order is reversed.

The research on language in apes forcefully illustrates a conflict, or tension, that is common to many other areas of research on learning in animals. If the investigators are interested in language and communication, they can attempt to communicate as naturally and informally as possible with their apes. This approach involves treating an ape as a fellow social being, with whom one plays and interacts as far as possible as one would with a human child; it also, almost inevitably, results in a style of research where it is exceptionally difficult to control precisely the cues that the ape may be using and even hard to avoid an overly rich, anthropomorphic interpretation of the ape's behaviour. If, on the other hand, the researchers are interested in rigorous experimental control and economical interpretation of the processes underlying the ape's performances, they are likely to set the ape formal problems to solve, with rewards for correct responses and no rewards for errors. But such an approach, however scientific it may seem, must run the risk of missing the point. This is not language; the investigators are not communicating with the ape in the way they would communicate with a child. The very nature of the experimental problems ensures that the ape will not use its language in the way that a child does: To communicate shared interests, to attract a parent's attention to what the child has seen or is doing, to comment on a matter of concern to both.

There is no resolution to this conflict, for both approaches have their virtues as well as their dangers, and both are therefore necessary. In just the same way, the study of a rat pressing a lever in a Skinner box or of a dog salivating to the ticking of a metronome seems to many critics a sterile and narrow approach to animal learning—one that simply misses the point that, if the ability to learn or profit from experience has evolved by natural selection, it must have done so in particular settings or environments because it paid the learner to learn something. It would be foolish to deny this obvious truism: of course it pays animals to learn. Indeed, it may pay them to learn quite particular things in specific situations, and different groups of animals may be particularly adapted to learning rather different things in similar situations. None of this should be forgotten, and the study of such questions requires the scientist to forsake the laboratory for the real world, where animals live and struggle to survive. But few sciences can afford to miss the opportunity to manipulate and experiment under laboratory conditions where this is possible, and none can afford to forget the benefits of precise observation under controlled conditions.

Animal Emotion

Emotions used to be an uncontroversial part of any description of animal behavior, as uncontroversial as the instincts with which they were compared and equated. William James—whose famous title "What is an Emotion?" inspired the one above— rightly regarded the emotions as an unlearned response system, which is precisely why the next century saw the study of emotions go out of fashion.

American behaviorism tried to explain all behavior based on operant conditioning and hence had no room for unlearned predispositions. Skinner dismissed the emotions as "excellent examples of the fictional causes to which we commonly attribute behavior." Until late into the last century, American researchers could scarcely obtain funding for work on the emotions unless they rephrased their questions in terms of learning and memory.

The second major behavioral school—European ethology—similarly abandoned anything considered as sentimental and imprecise as the emotions in a reaction against the subjective "animal psychology" of the time. Even to this day, the Oxford Companion to Animal Behaviour urges ethologists to avoid references to the emotions, because "It does nothing to promote our understanding of behavior to attribute it to an emotion if our only evidence of the emotion is the very behaviour the emotion is supposed to explain." Since the 1970s, ethology and its offshoots developed a strictly functionalist approach in which behavioral motivations barely counted, thus adding to the irrelevancy of the emotions. This functionalism went so far that Darwin's own pioneering comparisons between the emotional expressions of humans and other animals came to be regarded as un-Darwinian.

Despite the frequent assertion that animal emotions hardly matter, outright denial of their existence is rare. This leaves us with the curious situation that a widely recognized aspect of animal behavior is deliberately ignored or minimized. Emotions are often presented as too simple for attention. The Oxford Companion to Animal Behaviour asserts that "animals are restricted to just a few basic emotions," and the main difference between human and animal emotions has been proclaimed to be that "animals don't have mixed emotions." Whether animal emotions are pure and simple, however, cannot be ascertained without a scientific program to study them. One only needs to see an aroused chimpanzee, with all its hair on end, pick up a stick to safely poke at a snake that it has approached with great hesitation, to understand that mixed inclinations, such as between fear and curiosity, are entirely possible. Infact, when Menzel9 tested chimpanzees with toy snakes, he found that once one chimpanzee knew about the location of a snake, others who had never seen it would adopt the same cautious, ambivalent posture just from watching the first chimpanzee who had, thus showing the effectiveness and potential survival value of emotional communication.

Survival value is obvious to anyone who watches primate behavior, yet we seem unable to talk about emotions without putting them between skeptical quotation marks. We describe animals not as angry but as "angry" or aggressive, and not as loving, but as "loving" or affiliated and bonded. Greetings between animals may be called loud, elaborate, or intense, but rarely emotional. Apart from descriptive labels, functional labels are preferred provided they are devoid of intentionality. Thus, animals may be called "altruistic," but only in the functional sense in that they benefit others at a cost to themselves. The term is almost never used in its motivational sense to the perplexity of scholars outside behavioral biology, where altruism invariably implies benign feelings and intent. I experienced similar taboos when first describing how chimpanzees kiss and embrace their adversaries during reconciliations after a fight. That the term reconciliation is now widely accepted in primatology is the product of three decades of systematically countering "simpler" explanations, so that the only one left standing is that primates monitor the state of their social relationships and undertake reparatory actions following conflict.

Given how common reconciliation turns out to be among primates, as well as other social mammals, such as dolphins, hyenas, and goats, animals must have the capacity to substitute hostility with a friendly attitude, which in humans is ascribed to a complex emotional process known as forgiveness.

Do animals, too, know forgiveness? Even if this remains an unestablished fact, it seems prudent to keep an open mind, and replace Morgan's Canon— the traditional principle of cognitive parsimony— with a principle that better fits the evolutionary age. Instead of assuming that animal emotions are necessarily simple and straightforward, it is more likely that if humans and related species respond similarly under similar circumstances, the emotions behind their responses are similar, too. The latter view postulates fewer psychological changes in a relatively brief evolutionary time, hence is more parsimonious than the assumption that unique mechanisms underpin human behaviour.

This topic explores the concept of emotion through the eyes of a student of animal behavior and cognition. It is therefore light on neuroscience, focusing instead on the potential usefulness of emotional concepts for students of animal behavior. Neuroscience is present in the background, however, as the one discipline that can (and some would argue, has) cut the Gordian knot of enduring skepticism about animal emotions. Brain research may not be able to tell us what animals feel—even though it has revealed valences that must be centrally represented—but the argument from homology in the brain is immensely powerful. If humans report high anxiety while showing amygdala activation and rats exhibit flight and freezing responses when their amygdala is electrically stimulated, it is hard to avoid the conclusion that we are dealing with one and the same state, that is, fear. The same argument has been applied to emotional attachment, joy, anger, and so on, boosting the case for evolutionary continuity.

Emotions Potentiate Action

The antiquity of the emotions points at high adaptive value. A rise in core body temperature and increased heart rate not directly attributable to the eliciting stimulus—for example, mild handling by an experimenter—has been measured in mammals, reptiles, and birds, but not amphibians and fish. On the basis of this so-called emotional fever, it has been argued that the first elements of vertebrate anxiety emerged after the amphibians.

James married the emotions to the instincts. Both humans and other animals respond to danger with the emotion of fear, which is associated with the flight instinct. On the other hand, the thwarting of a goal causes frustration, which causes anger, also known as the aggressive instinct. A strange thing happened, though, when instincts went out of favor: The emotions remained, but in a seriously diluted form. Dunlap lamented: "Instincts have quietly passed away after a brief and feverish illness, and the widowed emotions have been left." For evolutionary biology, this dissociation was unfortunate as it removed the most likely reason why emotions exist, which is for their potential to induce adaptive action. What would be the point of reacting to the sight of a predator with a bodily state known as fear? Already one century ago we knew that this state is marked by deeper respiration, higher arterial pressure, a faster heart rate, a shifting of blood away from the digestive system to muscles, heart, and brain, increased vigilance, and the freeing of sugar from liver reserves. These changes by themselves do not do the organism any good: their adaptive value lies in bodily preparation for struggle or escape.

The beauty of the emotional response system, over an instinctual one, is that it is not strictly predetermined. The neurological and physiological changes it produces may be rapid and reflex-like, but the elicited behavior varies with situation and experience. Some primates have different alarm calls for different dangers to which listeners respond accordingly. The alarm call for big cats leads vervet monkeys to quickly climb a tree, the call for aerial predators makes them look up and run into dense bush, and the one for snakes makes them stand upright and look around in the grass. In all cases, the elicited emotion is fear, but it is an "intelligent" fear, one that seeks the most appropriate response to the circumstances.

This variability in response is important in relation to the question about the usefulness of the concept of emotion in behavioral analysis. The tautology charge against the concept, such as the one cited before from the Oxford Companion to Animal Behaviour, claims that a behavior indicative of a certain emotion cannot at the same time be explained by it. This charge loses much of its strength, however, if there is more than one behavioral indicator for a given emotion. It weakens even further if not only the outputs but also the inputs are variable. Fear, for example, may be triggered by a sudden noise, the sight or smell of a predator, a conditioned stimulus associated with an aversive event, an alarm call by a conspecific, or a rapidly approaching dominant individual. In humans, many subtypes of fear and anxiety deal with different kinds of threats.

The greater the range of inputs and outputs, the more useful it becomes to postulate a particular physiological and behavioral state that potentiates a response. When we say that individual A "fears" B, we are speaking of a different state with a different set of causes and consequences than when we say that individual A "loves" B. No precise future actions are predicted by these statements, but the behavioral likelihoods are vastly different. Behavioral states, or emotions, thus act as intervening variables in the same way that Hinde solved tautology charges against the concept of social dominance by proposing it as an intervening variable. The emotion concept ties correlated variables together, and economically summarize a wide range of inputs and outputs.

This view of emotions, as interfacing environmental challenges and optimal behavioral responses, is reflected in a variety of approaches. Barrett et al. describe an emotion as an orchestrated response to a significant event across multiple systems at once: perceptual, cognitive, motivational, expressive, bodily, and experiential. Cosmides and Tooby also stress the coordination aspect: "To behave functionally according to evolutionary standards, the mind's many subprograms need to be orchestrated so that their joint product at any given time is functionally coordinated, rather than cacophonous and self-defeating. This coordination is accomplished by a set of superordinate programs— the emotions." Frijda argues that the whole point of having emotions is goal-oriented action: "A passion wants something." However, this potential for action, which was central for McDougall and his contemporaries, is curiously absent from many recent definitions. Even if emotions happen inside the individual, they are triggered by the environment and predispose the organism's engagement with it. Their effect on behavioral outcomes is central to any evolutionary account, which assumes that emotions evolved to benefit the organism. In yet another variation on this theme, the philosopher Nussbaum attaches the Aristotelian concept of eudaimonism to the emotions in that they help the organism flourish.

The utilitarian perspective is a logical starting point to explain the existence of the emotions. Organisms have been selected to enter a particular bodily and mental state under particular circumstances: those who did furthered their interest better than those who did not. In the felicitous phrase of Lazarus and Lazarus, emotional reactions reflect "the wisdom of ages." My definition of an emotion incorporates these causal and functional insights:

An emotion is a temporary state brought about by biologically relevant external stimuli, whether aversive or attractive. The emotion is marked by specific changes in the organism's body and mind brain, hormones, muscles, viscera, heart, etcetera. Which emotion is triggered is often predictable by the situation in which the organism finds itself, and can further be inferred from behavioral changes and evolved communication signals. There exists no one-on-one relation between an emotion and ensuing behavior, however. Emotions combine with individual experience and cognitive assessment of the situation to prepare the organism for an optimal response.

Cognition and Emotion

Emotion research on humans has not always been as prominent and well accepted as it is today. It received a serious set-back during the cognitive revolution and its favorite comparison between mind and computer. In truly Cartesian fashion, neither the body nor the emotions were considered part of cognition, which was defined in terms of perception and memory.

Times have changed. The body is back in cognition research, and emotional intelligence is a topic of considerable interest. It is impossible to separate the emotions from cognition—attempts at which have been labeled a "sin" in the study of emotion. The brain has no separate cognitive and emotional pathways: what we attend to and the outcomes we are interested in is very much emotionally determined. The modern view is that the pathways overlap because of the need to coordinate processes and functions that are tightly linked. How else could the organism learn which stimuli to avoid or how to interpret the emotional signals of others? Even human rationality is far from emotion free: bodily states that accompany the emotions are part of its chain of operations. This integrative view goes back to David Hume, who went so far as to view reason as the slave of the emotions.

However, the field of animal cognition still shows a strong adherence to a disembodied mind. For example, instead of regarding imitation—originally defined as "doing an act from seeing it done"—as a learning process rooted in social closeness, bodily connections, and a desire to act like others, it was redefined top-down as a process requiring shared intentions. Tomasello and Call placed "true" imitation beyond the grasp of even apes, which "do not understand the other as an intentional agent who is similar to themselves as an intentional agent." Initially, this view was supported by negative findings on ape imitation, but the majority of studies employed human behavioral models behind a barrier of glass or mesh. In light of alternative theories of imitation, such a cross-species set-up is problematic, however. Alternative theories place less emphasis on intentionality and more on body-mapping and the neural fusion of perception and action, which process is likely enhanced by proximity and identification between observer and model. Once the extra effort had been made to train conspecific models and remove physical barriers, the issue of ape imitation was quickly settled to the point that major skeptics have come around. Chimpanzees faithfully and reliably copy tool use, foraging techniques, and arbitrary action sequences of species members. It was furthermore demonstrated that apes learn more from watching another ape open a puzzle box than from hundreds of disembodied demonstrations of the same box's mechanics.

Overly mentalistic approaches were also followed in the most popular research topic of the last few decades: theory of mind (ToM). Even though this topic came originally out of primate research, it is not widely realized that the very first studies about how one individual perceives the knowledge of another revolved around emotional body language. Unfortunately, the topic was soon redefined with a far more abstract focus, such as knowing what others know. The precise mechanism of such "mindreading" remained unaddressed, however, and to this date there is preciously little evidence, both for humans and other animals, in favor of the rational process implied by the word theory in theory of mind.

The acquisition of ToM probably starts with emotional connections. Children pass traditional ToM tasks around the age of 4 but appreciate the feelings, needs, and desires of others already at the age of 2 or 3. They often rely on emotional communication to deduce what kind of situation the other faces, showing reactions similar to Menzel's apes, who recognized if one among them had spotted hidden food or danger. It should not surprise, therefore, that after many studies in which apes were challenged to guess what human experimenters knew or did not know, the greatest research progress arrived when scientists adopted a more emotionally relevant approach by testing how one ape perceives the knowledge of another in a direct confrontation between dominant and subordinate individuals.

It is unlikely that perspective taking can be understood without close attention to emotional and bodily connections. In general, animal cognition research would do well to replace mentalistic approaches with a bottom-up perspective. We need to know more about the nuts and bolts of animal intelligence, including emotional factors. For example, just as emotional stimuli activate neural mechanisms that enhance human memory, it has been found that chimpanzees have a better recall of pictures of conspecifics in an emotional than neutral pose. We also know that chimpanzees, even though capable of following human gaze, are more interested in and motivated by their own species, and more willing to pay attention to their gaze, which may explain their poor performance compared to children when both are tested by humans on social cognition tasks. The lesson here is the same as for imitation and ToM research: Test paradigms should take emotional and bodily connections into account, which connections are by definition different within than between species.

Emotional Control

In discussions about the interaction between emotion and cognition, cognition is customarily assigned a superior position, as that which keeps emotions in check. Without cognition, so the argument goes, emotions would run rampant. Animals are often depicted as devoid of such controls, such as in one of the most prominent anthropological theories about the rise of human civilization, which assigns extraordinary weight to the taming of sexual urges. It seems more likely that human civilization is afforded by a long evolutionary history of neural mechanisms that regulate the emotions, such as those also known of other primates.

Sometimes, the relation between cognition and emotion is depicted as a battle between cortical and subcortical brain areas, reflecting a questionable hierarchical view of the brain. Given the two-way street between cerebral cortex and subcortical circuits, emotions are far from reflexive in that they include evaluations of the situation and the weighing of future actions. They are subject to powerful appraisal mechanisms inserted between stimulus and response, as explained by Scherer: "the special role of emotion seems to be that of an intelligent interface that mediates between input and output on the basis of what is most important to the organism at a particular time".

This decoupling of stimulus and response is quite adaptive in the hierarchical society of many primates. A young male chimpanzee may be visibly aroused by a sexually receptive female yet will need to find a way of mating with her out of view of dominant males, who may punish him. With cooperation of the female, he thus waits for the right occasion or engages in evasive tactics. Or, the alpha male may have received a pointed challenge from a younger male but first will do the rounds grooming his supporters before launching a counter attack later in the day. Or a female may experience the kidnapping of her infant by an inexperienced juvenile and is visibly distressed following the kidnapper around. She will need to get her offspring back without chasing the other up a tree or starting a fight, since both actions will be risky for her offspring. Often, mothers do attack the kidnapper, but only after they have retrieved their infant. And then there are the many cases of deception in which a chimpanzee will display a false emotion, such as when an older dominant female has not been able to get a hold of her younger opponent during a chase, after which she feigns conciliatory gestures to lure the other within reach in order to get even. All of these examples stem from Chimpanzee Politics.

Emotional inhibitions, even emotional camouflage, are of paramount importance in the complex societies of primates. That these animals are capable of impulse-control is supported by experiments on delayed gratification. Both apes and monkeys will pass up an immediate reward in favor of a better, delayed one. It has further been shown that chimpanzees, like children, play more with toys in the presence of accumulating rewards suggesting deliberate attempts at self-distraction in the face of temptation. Other studies have shown that apes can override an immediate drive in favor of future needs, an essential aspect of successful action planning. It seems, therefore, that the same intertwinement between emotion and cognition known of humans applies to our close relatives, including effective control over the emotions.

A Window on the Emotion

Facial displays are common in both primates and other visually oriented mammals, such as canids, felid and ungulates. These displays likely evolved in tandem with the ability to decode them so as to infer the emotional state of others through a perception-action mechanism that produces shared representations. A recent study even documented facial signs of pain in rodents. Research on facial expressions will open an evolutionary window on the emotions of animals the same way it has done for those of humans. The ability to communicate finely graded emotions is particularly striking in the hominids, with both apes and humans showing a rich repertoire of facial expressions owing to their highly differentiated facial musculature.

Darwin was the first to look at facial expressions the way, at the time, only a biologist would: namely, as structural features to be described and catalogued in the same way as plant or animal morphology. One of his objectives was to show how human facial expressions (a) constitute a shared heritage of our species, (b) have parallels with the expressions of other animals, and hence (c) provide a behavioral argument for evolutionary continuity. Humans may express joy and happiness differently than dogs, but all humans do it one way, and all dogs another way, indicating that the expression of emotions is species typical. For ethologists, these expressions constitute fixed action patterns.

The concept of homology is applied when the traits of different species can be traced back to a common ancestor. It is not unusual for homologous traits to vary in function, such as in the case of a bird's wing and the human arm, which both derive from the forelimb of the ancestor of birds and mammals. With regards to facial displays, homologous displays may have different meanings in different species through a motivational and functional recasting. Homology is usually contrasted with analogy, or convergent evolution, when similar traits (e.g., the fish-like shape of a dolphin) are independent products of similar environmental pressures. The distinction between homology and analogy is important in relation to fixed action patterns.

This evolutionary approach goes further than what Darwin proposed, but Darwin's strength was that he had picked the one feature of human behavior that seems to fit most or all of the above conceptualizations. In fact, facial expressions fit the mold of inborn behavior better than many of the behaviors now discussed as such in evolutionary psychology, such as mate selection. Not that these patterns necessarily lack a genetic component, but they are highly flexible and their occurrence varies with learning and environment. They are far removed from the stereotypical facial muscle coordinations and vocalizations, such as laughing and crying, that appear early in life and are remarkably uniform across individual humans and cultures. Not only did Darwin pick a prime candidate of innate behavior, he also recognized and carefully documented the similarity of our own facial movements with those of other primates. The implication was, of course, that if our own expressions bring emotions to the surface, then those of other primates probably do so as well.

With the notable exception of Ladygina-Kohts, who compared the expressive movements of her own son and a juvenile male chimpanzee, we had to wait until the 1960s for an extension of Darwin's observations. Van Hooff speculated about the causal underpinnings of facial expressions, such as the laugh and smile, tracing their phylogeny among a great variety of monkeys and apes. A recent acoustical analysis of the guttural panting by young apes being tickled confirmed the homology with human laughter to the point that a phylogenetic tree reconstructed from acoustic data matched a tree based on comparative genetics.

Van Hooff opted for a purely descriptive terminology followed by ways to establish the meaning of each display. A display associated with withdrawal would be considered fearful, for example, and one associated with attack would be considered aggressive. Those emotional interpretations would not affect the naming of the display, however. The problems the latter might cause can be illustrated with the common term fear grimace for the wide grin of macaques. This label obscures both the display's visual characteristics (a "grimace" being a mere facial contortion) as well as its precise meaning, which is closer to submission than fear, that is, the bared teeth display signals fear mixed with a desire for peaceful integration.

In addition to such overt expressions, primates often communicate by means of minor changes in the face. One of the most striking illustrations of this capacity is the cooperative avoidance paradigm in which one macaque needed to deduce from the face of another on a television screen if an electrical shock was forthcoming so as to turn it off in time. The monkeys were surprisingly effective at avoiding shock by extracting emotional information from their partner's face despite its apparent blankness, leading Miller to comment "It was our conviction that a monkey was a much more skilled interpreter of facial expression in another monkey than was man".

Emotional Contagion

Miller's experiment on facial cuing illustrates the value of emotional communication: one monkey reads another's subtle body language and reacts appropriately. But more may have been going on: not merely the reading of an emotion for one's own benefit, but actually adopting the same emotion as the other. Known as emotional contagion, this process forms the basis of empathy. Empathy relies on emotions: the capacity makes no sense without them.

Preston and de Waal propose that empathy rests on a perception-action mechanism that provides an observer (the subject) with access to the internal state and situation of another (the object) through the subject's own neural representations. When the subject attends to the object's state, the subject's neural representations of similar states that it has experienced are automatically and unconsciously activated. The more similar and socially close two individuals are, the easier this activation. This form of empathy may be widespread in mammals, such as the intensified pain response of mice that have watched other mice in pain.

Emotional contagion may lead individuals frightened by the alarm of others to hide or flee. It may cause a mother distressed by her offspring's vocalizations to reassure both herself and her offspring by warming or nursing them. It may inhibit an individual from inflicting pain upon another because of vicarious negative arousal triggered by the other's distress vocalizations. These empathic emotional reactions may benefit both the actor and individuals close to them. More complex forms of empathy occur when the subject takes the object's perspective, a capacity thought present in some large-brained mammals. Such perspective-taking permits targeted helping (i.e., helping geared toward to other's specific situation and needs) or the consolation of victims of aggression by means of kissing, embracing, grooming, and so on. The latter behavior has been extensively studied in apes, in which it follows predictions from the empathy hypothesis such as that consolation reduces the other's distress, is more common in females than males, and directed more to socially close than distant individuals.

Long before the current interest in animal empathy, McDougall recognized its importance for gregarious animals as "the cement that binds all animal societies together, renders the actions of all members of a group harmonious, and allows them to reap some of the prime advantages of social life." Because of their effect on others, emotions bring individuals together by converging internal states such as fear, hunger, playfulness, and sleepiness. The study of this phenomenon is crucial for our understanding of social life. This is as true for the spreading of alarm as it is for the role of the emotions in altruism and care. Empathy is considered the main proximate mechanism underlying human altruism, and the same has been argued for other mammals.

Animal Mimicry

The ability of an organism to imitate one or more traits from another organism so that it can obtain some benefit is known as mimicry. Mimicry occurs when one animal displays physical or behavioral traits that copy those of a different species or their surroundings, and incur a survival advantage on account of it.

All animal species try to increase their rate of survival and to achieve this goal in the most effective way is by looking similar to some other species. Mimicry is characterized by the superficial resemblance by two or more organisms which are not closely related taxonomically. It is not always between two or more organisms but it may also be when an organism displays behavioral or physical traits which resemble its surroundings. This phenomenon gives an advantage of protection from the predator to one or both organisms.

In the most studied examples, only one species gets the advantage by resembling the other (model) species. The phenomenon of mimicry in some butterflies was discovered in the 19th century which later included much more examples of plants and animals. In many cases, the organisms belonged to the same class, same order or even the same family. But the are many instances in which animals were mimicking some plants and vice versa. In most examples, the organisms were mimicking resemblance of some other organisms, however, some can also resemble in smell, sound or even biochemistry.

Types

There are many different types of it which can be classified into two main types: Defensive mimicry and Non-defensive mimicry.

Defensive Mimicry

Those species which have a lot of predators have adapted defensive mimicry because their survival rates are low if they do not avoid their predators. There are three subtypes of defensive mimicry.

Batesian Mimicry

In Batesian mimicry, the mimetic organism (usually harmless) copies the morphological traits like coloration and sounds of some poisonous or venomous organism in order to deceive the predators by making them think of it as harmful. So, the mimetic organism avoids being preyed by the predator. This phenomenon is known aposematism.

For example, a non-poisonous Scarlet King Snake uses the aposematic coloration of the poisonous Coral Snake to avoid its predators.

Müllerian Mimicry

In Müllerian mimicry, two or more types of poisonous and unpalatable species develop similar appearances in order to get the advantage of shared protection. In contrast to Batesian mimicry, Müllerian mimicry involves only poisonous and unpalatable species.

Different geographical forms of Heliconius erato and H. melpomene are the examples of Müllerian mimicry.

Mertensian Mimicry

This is a very unusual form of mimicry in which a deadly species mimics an aposematic trait of a less harmful species. How would it be beneficial?

This type of mimicry helps in lowering the predation rates because a predator eating a deadly species dies so it will be unable to transmit the information that the prey is harmful so the predation rate remains high. But if the deadly species resembles a less harmful species and the prey eats the less hamful species, it will transmit information to the other species that "Don't eat this prey, its harmful".

Non-Defensive Mimicry

This type is not used for defense but serves other purposes. The most important type of non-defensive mimicry is Aggressive or Peckamian Mimicry. Unlike defensive mimicry, in Peckamian mimicry, the predators develop the traits of a harmless or even beneficial species so they get unnoticed by their prey or hostage.

References

- Animal-behaviour, ecology-and-environmentalism, earth-and-environment: encyclopedia.com, Retrieved 5 March, 2019

- Ethology: newworldencyclopedia.org, Retrieved 15 June, 2019
- Animal-communication, science: britannica.com, Retrieved 7 April, 2019
- Animal, habitat: uen.org, Retrieved 19 July, 2019
- Types-of-adaptations-in-animals: animalwised.com, Retrieved 9 January, 2019
- Animal-learning, science: britannica.com, Retrieved 30 May, 2019

Chapter 5

Animal Diseases

There are a number of diseases which affect animals such as anthrax, blackleg, blue tongue, ovine brucellosis, exudative dermatitis, swine dysentery, caprine arthritis and caseous lymphadenitis. The chapter closely examines these diseases which affect different animals to provide an extensive understanding of the subject.

Animal disease is an impairment of the normal state of an animal that interrupts or modifies its vital functions. Concern with diseases that afflict animals dates from the earliest human contacts with animals and is reflected in early views of religion and magic. Diseases of animals remain a concern principally because of the economic losses they cause and the possible transmission of the causative agents to humans. The branch of medicine called veterinary medicine deals with the study, prevention, and treatment of diseases not only in domesticated animals but also in wild animals and in animals used in scientific research. The prevention, control, and eradication of diseases of economically important animals are agricultural concerns. Programs for the control of diseases communicable from animals to man, called zoo noses, especially those in pets and in wildlife, are closely related to human health. Further, the diseases of animals are of increasing importance, for a primary public-health problem throughout the world is animal-protein deficiency in the diet of humans. Indeed, both the United Nations Food and Agricultural Organization (FAO) and the World Health Organization (WHO) have been attempting to solve the problem of protein deficits in a world whose human population is rapidly expanding.

Detection and Diagnosis

Reactions of Tissue to Disease

Disease may be defined as an injurious deviation from a normal physiological state of an organism sufficient to produce overt signs, or symptoms. The deviation may be either an obvious organic change in the tissue composing an organ or a functional disturbance whose organic changes are not obvious. The severity of the changes that occur in cells and tissues subjected to injurious agents is dependent upon both the sensitivity of the tissue concerned and the nature and time course of the agent. A mildly injurious agent that is present for short periods of time may either have little effect or stimulate cells to increased activity. Strongly injurious agents in prolonged contact with cells cause characteristic changes in them by interfering with normal cell processes. Most causative agents of disease fall into the latter category.

Characteristics of Cell and Tissue Changes

Changes in cells and tissues as a result of disease include degenerative and infiltrative changes. Degenerative changes are characterized by the deterioration of cells or a tissue from a higher to a lower form, especially to a less functionally active form. When chemical changes occur in the tissue, the process is one of degeneration. When the changes involve the accumulation of materials within the cells comprising tissues, the process is called infiltration. Diseases such as pneumonia, metal poisoning, or septicemia (the persistence of disease-causing bacteria in the bloodstream) may cause the mildest type of degeneration—parenchymatous changes, or cloudy swelling of the cells; the cells first affected are the specialized cells of the liver and the kidney. Serious cellular damage may cause the uptake of water by cells (hydropic degeneration), which lose their structural features as they fill with water. The causes for the accumulation in cells of abnormal amounts of fats (fatty infiltration and degeneration) have not yet been established with certainty but probably involve fat metabolism. Poisons such as phosphorus may cause sudden increases in the accumulation of fats in the liver. An abnormal protein material may accumulate in connective-tissue components of small arteries as a result of chronic pneumonia, chronic bacterial infections, and prolonged antitoxin production (in horses); the condition is known as amyloid degeneration and infiltration. Hyaline degeneration, characterized by tissues that become clear and appear glasslike, usually occurs in connective-tissue components of small blood vessels as a result of conditions that may occur in kidney structures (glomeruli) of animals with nephritis or in lymph glands of animals with tuberculosis. Certain structures (glomeruli) of animals with nephritis result in degeneration.

The condition in which mucus, a secretion of mucous membranes lining the inside surfaces of organs, is produced in excess and accumulates in greater than normal amounts is referred to as mucoid degeneration. Major causes of this condition include chronic irritation of mucous membranes and certain mucus-producing tumours. Abnormal amounts of glycogen, which is the principal storage carbohydrate of animals, may occur in the liver as a result of certain inherited diseases of animals; the condition is known as glycogen infiltration. The abnormal deposition of calcium salts, which is known as hypercalcification, may occur as a result of several diseases involving the blood vessels and the heart, the urinary system, the gallbladder, and the bonelike tissue called cartilage. Pigments (coloured molecules) from coal dust or asbestos dust may infiltrate the lungs of certain dogs in two types of lung disease: anthracosis and asbestosis. Abnormal amounts of iron-containing coloured molecules (hemosiderin) resulting from the breakdown of hemoglobin, the oxygen-carrying protein of red blood cells, are often deposited in the liver and the spleen after diseases that involve excessive breakdown of red blood cells. A dark-coloured molecule (melanin) occurs abnormally in the livers of certain sheep suffering from Dubin–Johnson syndrome and in certain tumours called melanomas. Uric acid infiltration, which occurs in poultry, is characterized by the deposition of uric acid salts.

Necrosis, the death of cells or tissues, takes place if the blood supply to tissues is restricted; poisons produced by microbes, chemical poisons, and extreme heat or electricity also may cause necrosis. The rotting of the dead tissue is known as gangrene.

Atrophy of animal tissue involves a process of tissue wasting, in which a decrease occurs in the size or number of functional cells—e.g., in inherited muscular dystrophy of chickens. Hypertrophy—an increase in the size of the cells in a tissue or an organ—occurs in heart muscle during diseases involving the heart valves, in certain pneumonias, and in some diseases of the endocrine glands. Aplasia is the term used when an entire organ is missing from an animal; hypoplasia indicates arrested or incomplete development of an organ, and hyperplasia an increase in the production of the number of cells—e.g., the persistent callus that forms on the elbows of some dogs. Metaplasia is used to describe the change of one cell type into another; it may occur in chronic irritation of tissues and in certain cancerous tumours.

Characteristics of Inflammatory Reactions

When tissues are injured, they become inflamed. The inflammation may be acute, in which case the inflammatory processes are active, or chronic, in which case the processes occur slowly and new connective tissue is formed. The reaction of inflamed tissues is a combination of defensive and repair mechanisms. Acute inflammation is characterized by redness, heat, swelling, sensitivity, and impaired function. Several types of acute inflammation are known. Mild acute inflammations of mucous membranes resulting in the production of thin watery material (exudate) are called catarrhal inflammations; parenchymatous inflammations occur in organs undergoing degeneration. If the exudate formed in response to an injury is of a serious nature—that is, resembling blood plasma—the process is called serous inflammation. In fibrinous inflammation, a protein (fibrin) forms on membranes, including those in the lungs. In superlative inflammation, dead tissue is replaced with pus composed of colourless blood cells (leucocytes) and tissue juices.

During the inflammatory reaction, the injured tissue is surrounded by an area of rapidly dividing cells. Specialized cells called macrophages enter the tissue and remove blood and tissue debris. Other cells, called neutrophils, ingest disease-causing bacteria and other foreign material. In chronic inflammations, the connective tissue contains fibroblasts, cells that divide and form new connective, or scar, tissue.

Characteristics of Circulatory Disturbances

An increase in the rate of blood flow to a body part, which is referred to by the term congestion, or hyperemia, occurs during inflammation; a diminished blood flow to tissues is referred to by the term ischemia, or a local anemia. Examples of hemorrhage, the escape of blood from vessels, include epistaxis, or nosebleeds, in racehorses; hematemesis, or regurgitation of blood, in dogs with uremia; hemoptysis, or blood loss from lungs; hematuria, or blood in urine, of cattle with inflammation of the urinary bladder. Edema, a condition that is characterized by abnormal accumulations of fluid in tissues, occurs not only in a tissue during inflammation but also over the entire body if the concentration of blood-serum proteins, especially albumin, is low. A thrombosis, which is a blood clot in a blood vessel, may block or slow circulation of blood to tissues; if blood vessels become blocked, the condition is known as an embolism. The term infarction describes the necrosis that occurs in tissues whose blood supply is blocked by an embolism.

Methods of Examination

Before an unhealthy animal receives treatment, an attempt is made to diagnose the disease. Both clinical findings, which include symptoms that are obvious to a nonspecialist and clinical signs that can be appreciated only by a veterinarian, and laboratory test result, may be necessary to establish the cause of a disease. A clinical examination should indicate if the animal is in good physical condition, is eating adequately, is bright and alert, and is functioning in an apparently normal manner. Many disease processes are either inflammatory or result from tumours. Malignant tumours (e.g., melanomas in horses, squamous cell carcinomas in small animals) tend to spread rapidly and usually cause death. Other diseases cause the circulatory disturbances or the degenerative and infiltrative changes. If a specific diagnosis is not possible, the symptoms of the animal are treated.

A case record of the information pertaining to an animal (or to a herd of animals) that is suspected of having a disease is begun at the time the animal is taken to a veterinarian (or the veterinarian visits the animal) and is continued through treatment. It includes a description of the animal (age, species, sex, breed); the owner's report; the animal's history; a description of the preliminary examination; clinical findings resulting from an examination of body systems; results of specific laboratory tests; diagnosis regarding a specific cause for the disease (etiology); outlook (prognosis); treatment; case progress; termination; autopsy, if performed; and the utilization of scientific references, if applicable.

The veterinarian must diagnose a disease on the basis of a variety of examinations and tests, since he obviously cannot interrogate the animal. Methods used in the preparation of a diagnosis include inspection—a visual examination of the animal; palpation—the application of firm pressure with the fingers to tissues to determine characteristics such as abnormal shapes and possible tumours, the presence of pain, and tissue consistency; percussion—the application of a short, sharp blow to a tissue to provoke an audible response from body parts directly beneath; auscultation—the act of listening to sounds that are produced by the body during the performance of functions (e.g., breathing, intestinal movements); smells—the recognition of characteristic odours associated with certain diseases; and miscellaneous diagnostic procedures, such as eye examinations, the collection of urine, and heart, esophageal, and stomach studies.

General Inspection

Deviation of various characteristics from the normal, observation of which constitutes the general inspection of an animal, is a useful aid in diagnosing disease. The general inspection includes examination of appearance; behaviour; body condition; respiratory movements; state of skin, coat, and abdomen; and various common actions.

The appearance of an animal may be of diagnostic significance; small size in a pig may result from retardation of growth, which is caused by hog-cholera virus. Observation of the behaviour of an animal is of value in diagnosing neurological diseases; e.g., muscle spasms occur in lockjaw (tetanus) in dogs, nervousness and convulsions in dogs with distemper, dullness in horses with equine viral encephalitis, and excitement in animals suffering from lead poisoning. Subtle behavioral changes may not be noticeable. The general condition of the body is of value in diagnosing diseases that cause excessive leanness (emaciation), including certain cancers, or other chronic diseases, such as a deficiency in the output of the adrenal glands or tuberculosis. Defective teeth also may point to malnutrition and result in emaciation.

The respiratory movements of an animal are important diagnostic criteria; breathing is rapid in young animals, in small animals, and in animals whose body temperature is higher than normal. Specific respiratory movements are characteristic of certain diseases—e.g., certain movements in horses with heaves (emphysema) or the abdominal breathing of animals suffering from painful lung diseases. The appearance of the skin and hair may indicate dehydration by lack of pliability and lustre; or the presence of parasites such as lice, mites, or fleas; or the presence of ringworm infections and allergic reactions by the skin changes they cause. The poisoning of sheep by molybdenum in their hay may be diagnosed by the loss of colour in the wool of black sheep. Distension of the abdomen may indicate bloat in cattle or colic in horses.

Abnormal activities may have special diagnostic meaning to the veterinarian. Straining during urination is associated with bladder stones; increased frequency of urination is associated with kidney disease (nephritis), bladder infections, and a disease of the pituitary gland (diabetes insipidus). Excessive salivation and grinding of teeth may be caused by an abnormality in the mouth. Coughing is associated with pneumonia. Some diseases cause postural changes: for example, a horse with tetanus may stand in a stiff manner. An abnormal gait in an animal made to move may furnish evidence as to the cause of a disease, as louping ill in sheep.

Clinical Examination

Following the general inspection of an animal thought to have contracted a disease, a more thorough clinical examination is necessary, during which various features of the animal are studied. These include the visible mucous membranes (conjunctiva of the eye, nasal mucosa, inside surface of the mouth, and tongue); the eye itself; and such body surfaces as the ears, horns (if present), and limbs. In addition, the pulse rate and the temperature are measured.

The veterinarian examines the visible mucous membranes of the eye, nose, and mouth to determine if jaundice, hemorrhages, or anemia are present. The conjunctiva, or lining of the eye, may exhibit pus in pinkeye infections, have a yellow appearance in jaundice, or exhibit small hemorrhages in certain systemic diseases. Examination of the nose may reveal ulcers and vesicles (small sacs containing liquid), as in foot-and-mouth disease, a viral disease of cattle, or vesicular exanthema, a viral disease of swine. Ulceration of the tongue may be apparent in animals suffering from actinobacillosis, a disease of bacterial origin.

A detailed examination of the eye may show abnormalities of the cornea resulting from such diseases as infectious hepatitis in dogs, bovine catarrhal fever, and equine influenza. Cataract, a condition in which the passage of light through the lens of the eye is obstructed, may result from a disorder of carbohydrate metabolism (diabetes mellitus), infections, or a hereditary defect.

An elevated temperature, or fever, resulting from the multiplication of disease-causing organisms may be the earliest sign of disease. The increase in temperature activates the body mechanisms that are necessary to fight off foreign substances. Measuring the pulse rate is useful in determining the character of the heartbeat and of the circulatory system.

Tests as Diagnostic Aids

In many cases, the final diagnosis of an animal disease is dependent upon a laboratory test. Some involve measuring the amount of certain chemical constituents of the blood or body fluids, determining the presence of toxins (poisons), or examining the urine and feces. Other tests are designed to identify the causative agents of the disease. The removal and examination of tissue or other material from the body (biopsy) is used to diagnose the nature of abnormalities such as tumours. Specific skin tests are used to confirm the diagnoses of various diseases—e.g., tuberculosis and Johne's disease in cattle and glanders in horses.

Confirmation of the presence in the blood of abnormal quantities of certain constituents aids in diagnosing certain diseases. Abnormal levels of protein in the blood are associated with some cancers of the bone, such as multiple myeloma in horses and dogs. Animals with diabetes mellitus have a high level of the carbohydrate glucose and the steroid cholesterol in the blood. The combination of an increase in the blood level of cholesterol and a decrease in the level of iodine bound to protein indicates hypothyroidism (underactive thyroid gland). A low level of calcium in the serum component of blood confirms milk fever in lactating dairy cattle. An increase in the activities of certain enzymes (biological catalysts) in the blood indicates liver damage. An increase in the blood level of the bile constituent bilirubin is used as a diagnostic test for hemolytic crisis, a disease in which red blood cells are rapidly destroyed by organisms such as Babesia species in dogs and in cattle and Anaplasma species in cattle.

The examination of the formed elements of blood, including the oxygen-carrying red blood cells (erythrocytes), the white blood cells (neutrophils, eosinophils, basophils, lymphocytes, and monocytes), and the platelets, which function in blood coagulation, is helpful in diagnosing certain diseases. Examination of the blood cells of cattle may reveal abnormal lymphocytic cells characteristic of leukemia. Low numbers of leucocytes indicate the presence of viral diseases, such as hog cholera and infectious hepatitis in dogs. Neutrophil levels increase in chronic bacterial diseases, such as canine pneumonia and uterine infections in female animals. Elevated monocyte levels occur in chronic granulomatous diseases; e.g., histoplasmosis and tuberculosis. Canine parasitism and allergic skin disorders are characterized by elevated eosinophil levels. Prolonged clotting time may be associated with a deficiency of platelets.

Anemia has many causes. They include hemorrhages from blood loss after injuries; the destruction of red blood cells by the rickettsia Haemobartonella felis in cats; incompatible blood transfusions in dogs; the inadequate production of normal red blood cells, which occurs in iron or cobalt deficiency after exposure to radioactive substances; general malnutrition; and contact with substances that depress the activity of bone marrow.

Poisonings occur commonly in animals. Some species are more sensitive to certain poisons than others. Swine develop mercury poisoning if they eat too much grain that has been treated with mercury compounds to retard spoilage. Dogs may be poisoned by the arsenic found in pesticides or by strychnine, which is found in rat poison. Many plants are poisonous if eaten, such as bracken fern, which poisons cattle and horses, and ragwort, which contains a substance poisonous to the liver of cattle.

Examination of an animal's urine may reveal evidence of kidney diseases or diseases of the entire urinary system or a generalized systemic disease. The presence of protein in the urine of dogs indicates acute kidney disease (nephritis). Although constituents of bile normally are found in the urine of dogs, the quantity increases in dogs with the presence of infectious hepatitis, a disease of the liver. The presence of abnormal amounts of the simple carbohydrate glucose and of ketone bodies (organic compounds involved in metabolism) in an animal's urine is used to diagnose diabetes mellitus, a disease in which the pancreas cannot form adequate quantities of a substance (insulin) important in regulating carbohydrate metabolism. The urine of horses with azoturia (excessive quantities of nitrogen-containing compounds in the urine) or muscle breakdown may contain a dark-coloured molecule called myoglobin.

The presence of eggs or parts of worms in the excrement of animals suspected of suffering from intestinal parasites, such as roundworms, tapeworms, or flatworms, aids in diagnosis. Feces that are light in colour, have a rancid odour, contain fat, and are poorly formed may indicate the existence of a chronic disease of the pancreas. Clay-coloured fatty feces suggest obstruction of the bile duct, which conveys bile to the intestine during digestion.

The identification of a disease-causing microorganism within an animal enables the veterinarian to choose the best drug for therapy. Agglutination tests, which utilize serum samples of animals and microorganisms suspected of causing a disease, many times confirm the presence of the following bacterial diseases: brucellosis in cattle and swine, salmonellosis in swine, leptospirosis in cattle, and actinobacillosis in swine and cattle. Other tests measure the antibodies (specific proteins formed in response to a foreign substance in the body) formed against a disease-causing agent, such as those that cause brucellosis, foot-and-mouth disease, infectious hepatitis in dogs, and fowl pest.

The modern veterinary diagnostic laboratory performs, in addition to the tests mentioned, tests of cells in the bone marrow; specific-organ-function tests (liver, kidney, pancreas, thyroid, adrenal, and pituitary glands); radioisotope tests, tissue biopsies, and histochemical analyses; and tests concerning blood coagulation and body fluids.

Anthrax

Anthrax is a highly contagious and infectious soil-borne disease caused by Bacillus anthracis, relatively large spore-forming bacteria that can infect mammals. Anthrax is primarily a disease of herbivores, particularly bison and beef cattle.

Beef cattle and bison are most likely to contract the disease because they graze lower to the ground than many other herbivores, often due to drought or management practices. It is also believed that the abrasive forages they consume can injure the mouth and allow the bacteria quick access to the blood stream.

Three Types of Anthrax

1. Gastrointestinal: Caused when bacteria spores from an infected animal are ingested. The most common cause of anthrax in beef cattle.
2. Cutaneous: Contracted via a skin wound. The most common way that anthrax is contracted by people.
3. Pulmonary: Caused when the bacteria are inhaled. This type is the most dangerous, but is virtually impossible for animals to contract under natural conditions. This type of anthrax occurs in rare situations where people are working directly with contaminated hides.

When an animal dies of anthrax, the bacteria is present in most tissues of the body. If the tissues are not exposed to oxygen, the bacteria cannot form the spores that infect other animals, and quickly die off. However in most cases scavengers or carnivores open the carcass soon after death. Once the bacteria are exposed to oxygen, infectious spores begin to form.

Spores are tough, 'egg-like' microscopic structures that are difficult to destroy, and can survive for decades. Spores continue to live in the soil long after the animal has decomposed, especially in alkaline soils. Anthrax spores end up deep in the soil and are brought to the surface when the ground is disturbed, for example by:

- Digging (wells, ditches, pipelines, etc.),
- Heavy rains,
- Deep tilling,
- Overcrowded areas, or
- Soil erosion (e.g. wind, water, or wallows).

The spores brought to the surface can easily be ingested by cattle while grazing and cause infection. Spores concentrate in low spots when floodwaters evaporate and infect cattle that drink standing water. During droughts, animals graze closer to the ground and may consume soil. Contaminated feed and soil excavation can also spread anthrax.

Clinical Signs

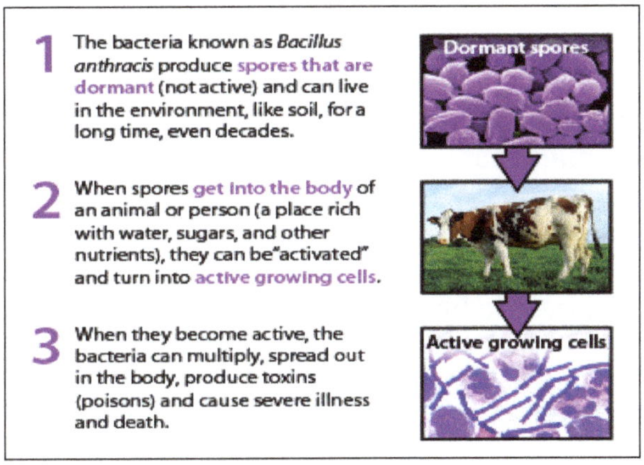

Anthrax in cattle has a very rapid onset. Once spores have been ingested, they infect macrophages (cells that are formed by the immune system in response to an infection), germinate and begin to multiply. The cells then produce a lethal toxin that kills body cells and causes excess fluid to accumulate in body tissues. Once the bacteria begin to multiply in the lymph nodes, the level of toxins in the body increase rapidly and cause tissue damage and organ failure. It is believed that the disease symptoms only last for 2-3 hours, so the most common symptom is sudden death. Very rarely observed symptoms are:

- Trembling,
- High temperature,
- Difficulty breathing,
- Convolutions and collapsing before death.

Often blood does not clot after death resulting in bloody discharge from anybody openings (rectum, mouth, nostrils, etc.)

Diagnosis

Anthrax is a federally reportable disease and in some provinces it is a provincially notifiable disease. Contact your veterinarian if you suspect anthrax on your farm. Veterinarians will collect blood samples to test for the anthrax bacteria. If anthrax is detected, it will be reported to the Canadian Food Inspection Agency. The provincial agriculture department may also be notified. The CFIA no longer investigates, tests, quarantines, vaccinates, or assists with anthrax mortality disposal, but some provinces may pay for diagnostic anthrax tests and provide advice on proper disposal and control.

Do not cut the carcass open. If anthrax is responsible, you want to keep the carcass intact to prevent blood leakage and exposure to air (which promotes anthrax spore formation). If the carcass has already been opened the spleen will be swollen, and there will be blood and fluid in the body cavities.

For any animal that dies suddenly and unexpectedly, cover the carcass with a tarp and call your veterinarian immediately. This will help prevent scavenging and disease spread.

Treatment

Anthrax is susceptible to most antibiotics, so prompt treatment of animals at the earliest sign of illness should be effective. Treatment with antibiotics does counteract the vaccine though, so they should not be provided to animals that were recently vaccinated.

Prevention

The best way to prevent the disease is by preventing the production and dispersal of spores.

- In consultation with your veterinarian, vaccinate cattle against anthrax if the disease has been a historical or current concern in your area, or the area you purchase forages.
- Manage grass so that it is not over grazed, taking particular care during drought conditions.
- Clean and disinfect any footwear or tools that may have come in contact with infected soils.
- Wash clothes that were worn when dealing with sick animals separately.
- Avoid using detergents with that contain calcium as disinfectants (e.g. use bleaches that contain sodium hypochlorite, not calcium hypochlorite) when anthrax is suspected, as calcium preserves spores.

Vaccination

An anthrax vaccine for beef cattle is licenced and available in Canada. After an initial vaccine and booster 2-3 weeks later, the Sterne vaccine provides immunity for approximately 1 year. Where acute risk of anthrax exists, vaccination is recommended three to four weeks before exposure.

Ask your veterinarian if vaccination is recommended in your herd. Vaccination is typically recommended if anthrax has historically occurred in an area or in an area a producer purchases hay from. Producers can obtain the vaccine through their veterinarian, who can order it from a supplier. The vaccine costs about $2.00 per dose.

While it is tempting to discontinue vaccination when cases have not occurred in the higher risk area for years, it is important to remember that spores can survive for decades and can easily resurface with changes in farming practices or weather patterns, so the risk is still present.

Although antibiotics such as penicillin can treat anthrax when used soon enough, it is important to not administer antibiotics within 8 days before or after vaccination, as they will interfere with immunity provided by the vaccine.

Disposal

Proper disposal helps limit the spread and recurrence of anthrax. The recommended process for disposal of cattle that may have contracted anthrax is a 3-step process:

1. Prevent blood or other fluids from leaving the carcass by plugging all body openings with an absorbent material (paper towel, cloth etc.), and covering the entire head with a heavy-duty plastic bag, that is secured by tape or tied securely.
2. Protect the carcass until disposal by covering it with a tarp, or other heavy plastic and weighting down the ends to prevent animals from getting in. This step is important even if you will be disposing of the carcass quickly, as it prevents anything from opening the carcass and spreading spores while setting up for disposal.
3. The recommended method of disposal is by incineration. Your veterinarian can tell you if incineration is legal in your area, recommend the best fuels to use, and the safest way to manage the incineration. When done properly, incineration completely destroys the bacteria and spores. Incineration should be done slowly so that all that remains is ashes and a few small pieces of bone. The ashes should then be soaked with an appropriate disinfectant (such as 10% formalin or 5% solution of lye (sodium hydroxide)) and buried using deep burial best practices.

If incineration is not an option, deep burial is also an acceptable practice, but should only be used when incineration is not a viable option. With burial, spores will still present and therefore carcasses should buried deep enough (6-8 feet) to prevent contaminated spores from being brought to the surface through weather or farming practices, yet 3 feet above the water table. Include the top layer of soil from where the animal was found dead as it is likely to be contaminated with spores. Do not bury carcasses in flood-prone areas.

If the carcass is not incinerated or buried, the anthrax bacteria will be exposed to air and form infectious spores as it decomposes. Dead stock pickup also increases the risk of future anthrax outbreaks by spreading the spores over a wider area. It is important to consult with your municipality or county about any permits that are required for incineration or burial, and record the coordinates of any burial or burn location on your farm.

When disposing an infected carcass, be sure to cover any open wounds you may have as it is possible to contract the cutaneous form of the disease if a break in your skin is exposed to the bacteria. Wear closed toe footwear, pants, long sleeve shirts and gloves when handling contaminated animals to minimize the risk of disease transmission, then wash clothing separately. Disinfect boots and wash hands and arms thoroughly. Anthrax does not naturally disperse in the air, so a breathing mask isn't necessary.

Blackleg

Blackleg is a fatal disease of young cattle. It produces an acute local infection, and the resulting blood poisoning leads to rapid death.

The name 'blackleg' derives from the fact that the site of infection is often a leg muscle, and that the affected muscle is dark in colour.

Occurrence of Blackleg

Although the disease is widely distributed in New South Wales, most losses from blackleg occur in the eastern half of the state. In some areas the disease may appear on several properties, while in other areas only isolated farms are affected. Sometimes, only part of a property is affected.

Susceptible Animals

Although blackleg has been found in cattle as young as 2 months old, most losses occur in cattle between 6 months and 2 years of age. Occasionally, losses are seen in adult cattle. Generally, the best conditioned animals are affected, with most losses occurring where there is an abundance of feed. Blackleg can occur at any time of the year, though more losses are seen during hot, humid weather or following the sudden onset of cold periods.

Cause

Blackleg is produced by spore-forming bacteria. The organisms most commonly responsible are Clostridium chauvoei and, less frequently, C. scepticism. Spores produced by the clostridia can lie dormant in the soil for years without losing their potency.

Method of Infection

Bacterial spores are eaten in contaminated feed or soil. The spores then enter the bloodstream and lodge in various organs and tissues, including muscles. Here they lie dormant until stimulated to multiply, possibly by some slight injury to the animal. The injury reduces blood flow to the area, thereby reducing the supply of oxygen to the tissues. In the absence of oxygen, the spores germinate and multiply. As they grow, the bacteria produce toxins which destroy surrounding tissues. The toxins are absorbed into the animal's bloodstream which makes the animal acutely sick and causes rapid death.

Signs of the Disease

Blackleg should be suspected if an animal aged between 6 months and 2 years:

- Becomes lame with swelling of a muscle;

- Stops grazing;
- Appears sick and quickly goes down.

However, these signs are usually of such short duration that they may be missed. More frequently, a thrifty calf or yearling is simply found dead. Gas is detectable under the skin and this produces a crackling sensation when the skin is rubbed with the hand.

The rapid accumulation of gas under the skin and in the body cavity gives the carcass a bloated appearance, with the limbs spread apart and pointing upwards. There may be a frothy, blood-stained discharge from the mouth, nostrils and anus. If the skin over the affected area is removed, excess bubbly blood-stained fluid can be seen, and the muscle immediately below will be dark in colour. However, when the affected muscle is inside the carcass, such as when the heart muscle is affected, no external evidence of the disease is found.

Figure shows diseased skeletal muscle from a heifer that died suddenly from blackleg. The muscle tissue is dark red and has a dry appearance due to gas formation. Because the disease is accompanied by rapid decomposition, the post-mortem changes may be masked by gas formation, especially if examination of the carcass is delayed for more than a few hours.

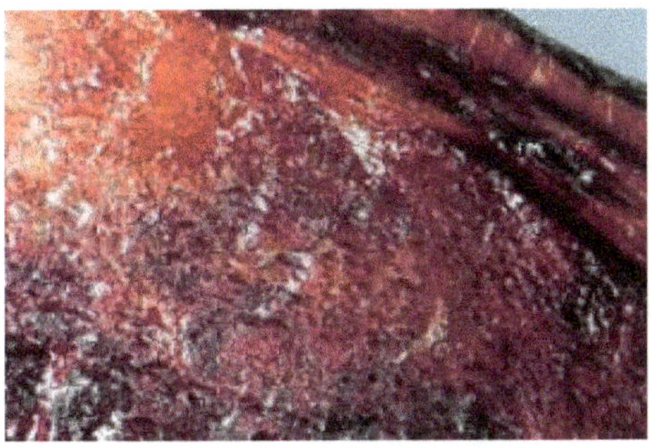

Figure: Diseased skeletal muscle.

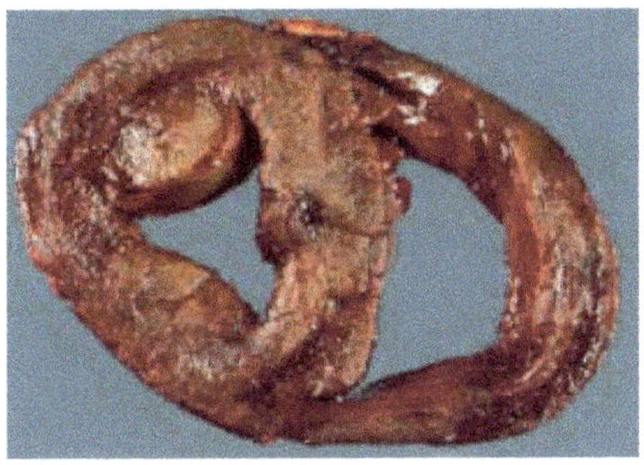

Figure: Diseased heart muscle.

Figure shows diseased heart muscle from a 12- month-old heifer that died suddenly from blackleg. Notice the haemorrhage and the 'half cooked' appearance of the diseased muscle tissue. The only effective means of controlling blackleg is by vaccination. Several makes of multivalent vaccine ('5 in 1' or '7 in 1') are available commercially and care should be taken to follow the manufacturer's instructions:

- Calves should receive two doses of blackleg vaccine. Two vaccinations 1 month apart are essential to provide the best protection.
- A booster vaccination 12 months later should provide lifelong immunity to blackleg.
- It is desirable to give the initial two doses of vaccine before young cattle reach their most susceptible age of six months.
- To await the occurrence of blackleg before vaccinating is unwise, as vaccines take 10– 14 days before they begin to provide immunity.

If store cattle are purchased in blackleg areas, it is wise to vaccinate all newly bought young cattle.

Vaccination Techniques

The vaccine should be delivered just under the skin – not into the muscle. Draw up a pinch of skin and insert the needle between the skin and the muscle. The loose skin of the neck is convenient for this. Do not save unused parts of bottles or containers of vaccines for future use, as they can become contaminated with undesirable organisms and/or lose their potency. Destroy any vaccine not used within 24 hours of opening.

Vaccination Breakdowns

Modern vaccines are produced under conditions of strict quality control by reputable manufacturers. Occasionally, reports are received of apparent failure of vaccines. When investigated, most of these alleged vaccination failures are due to:

- Incorrect dosing;
- Faulty technique;
- Using time-expired vaccine;
- Using material from partly used containers;
- Vaccine having been subjected to high temperatures during storage or transportation.

Abscesses or large swellings at the site of vaccination are usually the result of:

- Lack of hygiene;
- Injecting into muscle rather than beneath the skin.

If all these factors can be ruled out, and breakdown is suspected, seek veterinary advice.

Treatment

The speed with which blackleg kills usually makes individual treatment useless. In some cases, however, animals treated early with antibiotics may survive, although they often suffer permanent deformity due to partial or complete destruction of muscles.

Carcass Disposal

Carcasses of animals known to have died from blackleg should not be opened. Opening the carcass can liberate bacteria which will form spores that will contaminate the ground and subsequently infect other cattle. Also, do not drag carcasses along the ground. If possible, burn or deeply bury the carcasses where they lie.

Blue Tongue

Bluetongue is a viral disease of ruminants transmitted by midges in the genus Culicoides. Bluetongue virus is very diverse: there are more than two dozen serotypes, and viruses can reassort to form new variants. This virus is endemic in a broad, worldwide band of tropical and subtropical regions from approximately 35°S to 40°N; however, outbreaks also occur outside this area, and the virus may persist long-term if the climate and vectors are suitable. While overwintering in regions with cold winters is unusual, bluetongue virus recently demonstrated the ability to survive from year to year in central and northern Europe.

Bluetongue virus can replicate in many species of ruminants, often asymptomatically. Clinical cases tend to occur mainly in sheep, but cattle, goats, South American camelids, wild or zoo ruminants, farmed cervids and some carnivores are occasionally affected. Cases range in severity from mild to rapidly fatal, and animals that survive may be debilitated. Additional economic costs result from reproductive losses, damaged wool and decreased milk production. Control of this vector-borne disease is difficult, except by vaccination. The existence of multiple serotypes complicates control, as immunity to one serotype may not be cross protective against others.

Etiology

Bluetongue results from infection by bluetongue virus, a member of the genus Orbivirus and family Reoviridae. At least 26 serotypes have been identified worldwide. A few bluetongue viruses have additional names (e.g., Toggenburg orbivirus for the prototype strain of serotype 25). Isolates differ in virulence, and some strains seem to cause few clinical signs. Like some other viruses such as influenza virus, bluetongue viruses can reassort and recombine to produce new variants.

Bluetongue viruses are closely related to the viruses in the epizootic hemorrhagic disease (EHD) serogroup, a factor that can influence the development and/or selection of some diagnostic tests.

Species Affected

Bluetongue virus can infect many domesticated and wild ruminants including sheep, goats, cattle, water buffalo, African buffalo (Syncerus caffer), bison (Bison spp.), various cervids, wild relatives of sheep and goats, wildebeest (Connochaetes spp.) and other species. Both domesticated ruminants (especially cattle) and wild ruminants can be maintenance hosts. The virus can also infect camelids, and antibodies have been detected in some nonruminant wildlife including African elephant (Loxodonta africana), black and white rhinoceros (Diceros bicornis and Ceratotherium simum) and giraffe (Giraffa camelopardalis) in Africa, and collared peccaries (Pecari tajacu) in South America.

Among domesticated animals, clinical cases mainly occur in sheep; however, cattle, goats, yaks (Bos grunniens), llamas and alpacas were also affected in some outbreaks. In North America, cases of bluetongue have been documented in wild white-tailed deer (Odocoileus virginianus), pronghorn (Antilocapra americana) and bighorn sheep (Ovis canadensis), and captive Reeve's muntjac (Muntiacus reevesi) and greater kudu (Tragelaphus cupensis). There are no published reports of outbreaks among wild ruminants in Africa or Europe; however, clinical cases were reported in some zoo animals during serotype 8 outbreaks in Europe. Affected species included North American bison (Bison bison), European bison/ wisents (Bison bonasus), yaks, musk ox (Ovibos moschatus), Alpine ibex (Capra ibex), Siberian ibex (Capra sibirica) mouflon (Ovis aries musimon), blackbuck (Antilope cervicapra), fallow deer (Dama dama) and a Bactrian camel (Camelus bactrianus).

Accumulating evidence suggests that carnivores can also be infected by bluetongue virus. Antibodies to this virus have been detected in dogs, cats, cheetahs (Acinonyx jubutus), lions (Panthera leo), wild dogs (Lycaon pictus), jackals (Canis spp.) spotted hyenas (Crocuta crocuta) and large-spotted genets (Genetta maculata). Clinical signs have been reported in pregnant dogs infected by serotype 11, and nonpregnant Eurasian lynx (Lynx lynx) infected by serotype 8. Bluetongue virus is not zoonotic.

Geographic Distribution

Bluetongue virus can be found worldwide within tropical and subtropical climates from approximately 35° S to 40° N, and in some areas outside this region (e.g., in parts of California). Endemic areas exist in Africa, Europe, the Middle East, North and South America and Asia, as well as on numerous islands (e.g., Australia, the South Pacific, the Caribbean). Multiple serotypes can be found in many regions.

Outbreaks can occur outside endemic areas, but in most cases, the virus does not persist once cold weather kills the Culicoides vectors. Unusually, a serotype 8 virus overwintered for multiple years in central and northern Europe.

Transmission

Bluetongue virus is mainly transmitted by biting midges in the genus Culicoides, which are biological vectors. These midges can fly short distances of 1- 2 km, but they can be blown much farther by wind. Some species known to be effective vectors include Culicoides sonorensis in the United States; C. brevitarsis in Australia and parts of Asia; C. imicola in Africa, the Middle East, southern Europe and parts of Asia; C. bolitinos in some cooler regions of Africa; and C. insignis in the Caribbean, Central and South America and parts of the U.S. Additional species (e.g., members of the C. obsoletus-dewulfi complex in northern Europe) may be important locally. Other biting arthropods such as sheep keds (Melophagus ovinus), cattle lice (Haematopinus eurysternus), ticks and mosquitoes might be capable of transmitting the virus mechanically, but their role, if any, is thought to be minor. Bluetongue virus can also be spread mechanically on surgical equipment and needles. Some field strains and attenuated vaccine strains can infect the fetus in utero. Some of these animals can be born infected, and may introduce the virus to new areas if the dam is transported. Bluetongue virus can overwinter in some cold regions, by unknown mechanisms.

Bluetongue virus can persist in the blood of some animals for relatively long periods, facilitating transmission to Culicoides. Live virus has been isolated from some cattle for as long as 5 to 9 weeks, and viral RNA has been found much longer. Prolonged viremia has also been reported in other species, such as red deer (Cervus elaphus). However, animals eventually clear the virus, and there is no evidence that they remain persistently infected, even when infected in utero. Serotype 25 (Toggenburg orbivirus) in goats might be an exception. In one report, serotype 25 viral RNA was found for at least 2 years in individual animals, and the blood of some goats was infectious at 12-19 months.

At least some bluetongue strains (including members of serotypes 1, 8 and 26) can be transmitted directly between ruminants in close contact. This is a newly recognized route, and is generally thought to be of little epidemiological significance compared to transmission by midges. Serotypes 25 and 26 might be exceptions, as these viruses do not seem to replicate readily in some Culicoides vectors or in cell lines derived from these midges. The mechanisms of contact transmission are still uncertain. Suggestions have included shared feed and water troughs, contamination of wounds with blood during fighting among red deer, and contact with infected placentas in a case in cattle. Oral transmission of bluetongue virus has been demonstrated in colostrum, and experimentally via cultured viruses or urine from an experimentally infected sheep. Serotype 26 nucleic acids were found at low levels in nasal and ocular secretions of goats, and this virus was isolated from ocular swabs. Bluetongue virus can also be shed in semen.

How bluetongue virus infects carnivores is still uncertain. Seropositive dogs, cats and wild carnivores in Africa were thought to have eaten tissues from infected animals, and Eurasian lynx in a zoo had been fed ruminant fetuses and stillborn animals from outbreak areas. However, seropositive dogs in Morocco had not been fed raw diets, suggesting that they were infected by Culicoides. Bites from midges were also thought to have been responsible for some clinical cases among pregnant dogs in the U.S., although other routes (e.g., infection via semen or a contaminated semen extender) could not be entirely ruled out. Other dogs became infected when they accidentally received a bluetongue virus-contaminated vaccine. Transplacental transmission has been demonstrated in dogs, at least for serotype.

Disinfection

Disinfectants reported to be effective against bluetongue virus include sodium hypochlorite and 3% sodium hydroxide. Sodium hypochlorite, iodine (potassium tetraglicine triiodide) and a quaternary ammonium disinfectant (didecyldimethylammonium chloride) could inactivate another orbivirus, African horse sickness virus, and may also be active against bluetongue. The incubation period is estimated to be approximately a week, with a range of 2-10 days.

Clinical Signs

Clinical cases of bluetongue occur mainly in sheep, while subclinical infections seem to predominate in most other species.

Sheep

Sheep infected with bluetongue virus may remain asymptomatic, or become mildly to severely ill. Common clinical signs include fever and depression; serous to mucopurulent nasal discharge, which may crust around the nostrils; and hyperemia of the muzzle, oral and nasal mucous membranes, conjunctiva and coronary band of the hooves. The muzzle, periocular region and face often become edematous; in some cases the swelling may also involve the ear and/or submandibular region, and occasionally even extends as far as the axillae. The lips and tongue may be very swollen in some animals; the tongue is occasionally cyanotic in severe cases, and may protrude from the mouth. Petechiae and ecchymoses can also develop on the muzzle, oral mucous membranes and coronary band. The oral lesions, which often include erosions and ulcerations, can result in drooling, soreness during eating, or anorexia. Involvement of the coronary bands leads to lameness, with hot and painful hooves; sloughing of the hooves is possible. The udder and teats may also have lesions, and muscle damage can result in torticollis. Some sheep with bluetongue develop pulmonary edema and die rapidly, with clinical signs of dyspnea.

Pregnant ewes can abort or give birth to lambs that are stillborn or have CNS lesions, retinal lesions and/or skeletal malformations. CNS lesions can result in neurological signs or "dummy" lambs that cannot nurse or follow the ewe. The specific syndromes vary with the stage of gestation, and lambs infected later in the pregnancy can be born normal.

Deaths are often the result of pulmonary edema in acute cases, or secondary bacterial complications and exhaustion when the course is more prolonged. In animals that survive, sequelae may include general loss of condition and hoof deformities. Some surviving sheep have abnormal wool growth, or lose some or all of their wool a few weeks after the illness, and poor quality semen has been reported transiently in rams. Mildly affected sheep usually recover rapidly.

Wild Relatives of Sheep and Goats

Some bighorn sheep in North America and European mouflon affected by serotype 8 outbreaks in Europe had clinical signs resembling classical bluetongue in sheep, while other animals died suddenly without preceding signs. Hemorrhages were also reported. Nasal discharge and sudden death were seen in Alpine ibex in European zoos during the serotype 8 outbreaks. A Siberian ibex developed swelling of the head and neck, but survived.

Cattle and Goats

Infections in cattle and goats are usually subclinical in endemic areas. These species are more likely to be affected when a naive population is first exposed to bluetongue virus, such as during recent serotype 8 outbreaks in Europe. Clinical cases resemble the disease in sheep, but tend to be milder. Reported signs in cattle include in appetence, lethargy, facial edema, submandibular edema, oral inflammation with vesicles or ulcers in the mouth, excess salivation, nasal discharge, an elevated respiratory rate, edema of the distal limbs, and hyperemia of the coronary band with lameness. The muzzle of some cattle was reported to have a "burned" cracked appearance. Body temperature was sometimes normal. Cattle can also develop various skin lesions including vesicular and ulcerative dermatitis, periocular dermatitis, necrotic lesions, sloughing of affected skin and photodermatitis. In some cases, the skin may develop thick folds and cracks, particularly around the withers and neck. Udder and teat lesions, such as erythema, ulcers, cracking and necrotic lesions, have been reported in both cattle and goats, and milk production can be decreased. In some goats, a high fever and drop in milk production were reported to be the only sign. Abortions and stillbirths have been reported in both species, and congenital abnormalities including CNS lesions have been observed in newborn calves. Temporary sterility may be seen in bulls. Deaths are possible, but uncommon.

Post Mortem Lesions

In addition to the external lesions seen in living animals, such as facial edema and coronary band lesions, sheep may have hyperemia, hemorrhages, erosions and ulcers in the mucosa of the gastrointestinal tract from the mouth to the forestomachs, particularly where mechanical abrasion occurs (e.g., the buccal surface of the cheek and the mucosa of the esophageal groove and omasal fold). The heart may contain petechiae, ecchymoses and necrotic foci; in bluetongue, focal necrosis is particularly common in the papillary muscle of the left ventricle. Subintimal hemorrhage at the base of the pulmonary artery is also characteristic. Pulmonary edema, which may be accompanied by pleural and pericardial effusion, is a common cause of death in fatal cases, and may be a prominent finding. Hyperemia, hemorrhages and/or edema may also be detected in other internal organs. In addition, the skeletal muscles may have focal hemorrhages or necrosis, and the intermuscular fascial planes may be expanded by edema fluid. Lesions in fetuses and newborn lambs can include cavitating lesions in the brain, hydranencephaly, porencephaly, retinal dysplasia and skeletal abnormalities.

Similar signs may be seen in other species. Pulmonary edema was the most prominent finding in llamas, alpacas, and experimentally infected pregnant dogs. Widespread petechial to ecchymotic hemorrhages were common in some animals, including some white-tailed deer. A Eurasian lynx that died with virologically confirmed bluetongue virus infection had gross lesions of petechial hemorrhages, subcutaneous hematomas, anemia and lung congestion with edema. A second lynx that died several months later was emaciated, with anemia, enlarged and gelatinous lymph nodes, petechial hemorrhages and pneumonia.

Diagnostic Tests

Bluetongue virus can be found in blood from living animals, and in spleen, lymph node or bone marrow samples collected at necropsy. Reverse transcriptase-polymerase chain reaction (RT-PCR) tests are widely used to identify viral RNA in clinical samples, and can also identify the serotype. Bluetongue virus can be isolated in embryonated chicken eggs or various mammalian or insect cell lines (e.g., KC [Culicoides variipennis] cells). Animal inoculation in sheep or suckling mice is not usually employed, except where cell culture facilities are unavailable; however, it is reported to be more sensitive than isolation in cell culture. Blood samples for virus isolation should be collected as early as possible after infection. Bluetongue viruses are usually identified (i.e., to the bluetongue serogroup level) by RTPCR. Immunofluorescence/immunoperoxidase staining or group-specific antigen-capture ELISAs may also be used. Viruses can be serotyped by RT-PCR, gene sequencing or virus neutralization tests. Serotyping by virus neutralization can be difficult to interpret, as some serotypes cross-react in this assay.

Serology can be used to identify animals that have been infected or exposed to bluetongue virus. Antibodies typically appear 7 to 14 days after infection and are usually persistent. ELISAs are often used, but agar gel immunodiffusion (AGID) or virus neutralization can also be employed. AGID cannot reliably distinguish whether the animal was infected by bluetongue virus or epizootic hemorrhagic disease (EHD) virus; however, monoclonal antibody-based competitive ELISAs can distinguish antibodies to these two viruses. An indirect ELISA can detect antibodies to bluetongue virus in bulk milk samples. Complement fixation has largely been replaced by other tests, although it may still be used in some countries. No specific treatment is available, other than supportive care.

Control

Disease Reporting

Bluetongue is difficult to control once it has been transmitted to its vectors, and infections should be reported quickly in countries where this virus is not endemic. Reporting requirements in endemic areas may vary, depending on factors such as the existence of a control program. Appropriate sources (e.g., state authorities in the U.S.) should be consulted for the most current information.

Prevention

Bluetongue is mainly controlled by vaccination. Vaccines should be matched to the viral serotype; protection against other serotypes can be limited or nonexistent. Both attenuated and killed vaccines are currently made (although not necessarily sold in all regions), and multivalent vaccines are available. Attenuated vaccines are considered to be more effective than killed vaccines; however, midges may transmit these vaccine strains to unvaccinated animals during the vector season. Such vaccine strains could reassort with field strains, although how often this happens is currently unclear. Attenuated vaccine strains can also cause fetal malformations in pregnant ewes, and in some cases, may be able to cause systemic disease in highly susceptible animals.

Surveillance of sentinel animals may detect bluetongue viruses before outbreaks occur, and allow vaccination campaigns or other controls to be implemented early. Movement controls for infected animals (including pregnant, seropositive animals) may help limit virus introduction into new areas. Measures to reduce exposure to the Culicoides vectors may also be helpful during outbreaks or in endemic regions, although they are unlikely to be effective as the sole control measure. Such measures can include avoidance of environments where midges are more prevalent (e.g., low-lying, damp pastures), stabling animals from dusk to dawn, and/or the use of insecticides or insect repellents (e.g., insecticide-impregnated nets in stables) to help protect groups of animals. Effective vector control is challenging, due to factors such as the extensive breeding sites and large populations of Culicoides, and there are also environmental concerns with widespread use of pesticides. Some species of Culicoides are now known to enter barns and stables, especially late in the season when temperatures are becoming colder. Direct contact transmission is thought to have only a minor role in most outbreaks; however, disinfection and other infection control measures might be considered in some situations.

Morbidity and Mortality

Bluetongue is a seasonal disease in many areas, due to fluctuations in vector populations caused by cold temperatures or other factors such as rainfall. In regions where multiple serotypes circulate, the dominant serotypes may differ between years. Seroprevalence rates vary widely in endemic regions, ranging from 1% to more than 80% in domesticated and wild ruminants such as sheep, goats, cattle and cervids, as well as in some other species such as camels. In Europe, red deer seem to be the most important wildlife species; during serotype 8 outbreaks, more than half the red deer surveyed in some areas were seropositive. Some studies in endemic areas have not detected antibodies in South American camelids; however, 14% of these animals were seropositive during serotype 8 outbreaks in Germany, with rates varying from <1% to 43% in different parts of the country. Infections in carnivores are still poorly understood, but antibodies were found in 21% of dogs sampled in an endemic region of Morocco, as well as in some dogs, cats and wild carnivores in Africa. A few seropositive dogs have been reported in the U.S.

Sheep are usually the most severely affected species, and in endemic regions, they are often the only domesticated animal with obvious clinical signs. Morbidity rates in sheep range from < 5% to 50-75% (or higher), and are usually at their highest when the virus is first introduced. The case fatality rate is typically < 30%, but can reach 50-90% in highly susceptible populations. Once a virus has become endemic, morbidity may decrease to low levels (e.g., 1-2%), with very few deaths. Outbreaks are uncommon where a virus circulates year-round. Other species such as cattle and goats can also be affected when a bluetongue virus is introduced into a naive population. Many cattle became ill during recent serotype 8 outbreaks in Europe. Generally, cases in cattle and goats are milder than in sheep, and the mortality rate is lower. However, the case fatality rate was reported to be 26% in some goats infected with serotype 8 in Germany. Few cases have been documented in llamas and alpacas, and they were generally fatal. Among wildlife, whitetail deer and pronghorn antelope can be severely affected, with morbidity rates reported to be as high as 100% and case fatality rates up to 80-90%. Disease severity can also be influenced by factors such as the virus strain, host factors (e.g., immunity, general health, genetic factors, age and possibly breed) and environmental stressors.

Ovine Brucellosis

Ovine brucellosis is a reproductive disease that can affect all breeds of sheep. The disease may cause economic loss in flocks by reducing lambing percentages, extending lambing seasons and increasing ram culling. Once established in a flock, repeated blood testing and ram culling is required to eradicate the disease. Ram breeders in Western Australia can participate in a voluntary Ovine Brucellosis Accreditation Scheme.

Ovine brucellosis is an infection of sheep caused by a bacterium, Brucella ovis (B. ovis). The disease occurs in most sheep-producing areas of the world. Ovine brucellosis occurs naturally only in sheep. It is a disease primarily of rams. Ewes usually only carry the infection for one or two oestrus cycles.

In rams the disease causes inflammation of the epididymis (the vessel which transports semen from the testes). One or both ram testes may be affected. In pregnant ewes, B. ovis causes inflammation of the placenta.

Signs in Rams

Initial infection by B. ovis causes an elevated temperature and depressed appetite, which usually pass unnoticed by owners. The infection first affects the epididymis causing inflammation and swelling in the surrounding tissues. The epididymides may become completely blocked, causing sperm to build up. Obvious swelling, hardening or differences in size can often be seen and palpated through the scrotum. Rams with chronic infection may have a grossly enlarged tail of the epididymis at the lower end of the testis and a shrunken testis.

In some rams, the blockage is permanent and the tail of the epididymis continues to swell. In other rams the blockage may break down, the swelling reduce and the testes feel normal, while the ram may still be infected. Other causes of epididymitis and scrotal lesions include infections with other bacteria, such as Actinobacillus seminis and Corynebacterium pyogenes, and physical injury. Palpate scrotal contents of all rams to help determine soundness before mating.

Signs in Ewes

Clinical signs are not common in ewes. Most ewes will not remain infected for more than two oestrus cycles after being served by an infected ram. Infection causes inflammation of placenta, early embryonic death usually without outward signs or uncommonly the foetus may be aborted. The brucella organism may be present in vaginal discharges and milk, after abortion or parturition. Infection in ewes is self-limiting, so ewes usually breed normally in subsequent years.

Effect on Flock Fertility

Reduced Lamb Marking Percentages

When clinically affected rams are used, the lambing percentages may be reduced. If brucellosis continues in the flock, lambing percentages will progressively reduce.

Extended Lambing Period

When infected rams service ewes, the chance of conception is low and many ewes will return to service 17 days later. If rams are left with ewes, the number of pregnant ewes will increase, but this will result in an unusually long lambing period.

High Ram Culling Rate

If rams are culled when abnormalities are detected on palpation, then infected rams are culled one or two years after infection. If brucellosis is not eliminated, this high culling rate will continue.

Spread of Infection

Rams are infected by exposure to infected semen or vaginal discharges from infected or aborting ewes. B. ovis can be spread within a flock through:

- Rams joining ewes that have been served by infected rams in the same heat period.
- Homosexual activity, common among young rams.
- Rams exposed to infected ewes that have aborted and are excreting B. ovis infection through the nasal passage due to sniffing and nosing behaviour of rams in ewe flocks rams running with killer mobs or culls. Several rams may serve any ewe on heat.

Introduction of B. ovis into a previously clean flock is through buying, borrowing or straying infected rams.

Diagnosis

There are three methods of diagnosing ovine brucellosis: palpation of the scrotum, examination of semen and blood testing.

Palpation of the Scrotum Contents (Testes and Epididymes)

Examine ram in a standing or sitting (shearing) position. Use both hands and start above the testes, moving down to the base of the testes and the epididymal tails. Check for symmetry, free movement in the scrotum and the absence of lumps within the testes and epididymides. Testes should feel firm and springy.

Abnormalities will usually be felt as a swelling below either or both testes, due to inflammation of the tail of the epididymis. There may be a swelling at the head of the epididymis, above the testes. Not all brucella-infected rams will have palpable abnormalities, and conversely, rams with detectable abnormalities may not have brucellosis.

Semen Examination

Examination of semen may reveal a reduced volume of semen, poor sperm motility and/or a high proportion of abnormal sperm. B. ovis may also be cultured from semen.

Blood Tests

Blood is tested to detect the presence of B. ovis antibodies, which indicates previous exposure to infection.

Eradicating Ovine Brucellosis from Breeding Flocks

When ovine brucellosis is detected in a flock, the owner should discuss the management options with their veterinarian. Eradication can be achieved by a combination of manual examination, blood testing, and removal of infected rams as soon as they are detected. There is no effective vaccine. During an eradication program it is important to prevent contact between infected and uninfected young rams. Eradication is not compulsory. Eradication can be a slow and expensive process, and it may fail if testing, segregation and culling are not practised carefully. Once eradication is achieved, precautions must be taken to prevent infected rams re-entering the flock, whether as strays or introductions.

Infected Rams

Separate infected rams immediately from other tested-negative and young rams. Sell infected rams for slaughter.

Treatment is rarely undertaken and likely to be economical only in valuable rams. Prolonged antibiotic treatment may be effective in early cases before irreparable epididymis damage occurs. Unilateral castration is unlikely to be successful as the bacteria may also be present in the other epididymis and the internal accessory sex glands.

Prevention

Owners must ensure that introduced rams are from a flock with a low risk of having ovine brucellosis. A single negative blood test prior to purchase can be misleading because antibodies are not detected in the blood for up to seven weeks after a ram is infected.

Exudative Dermatitis

Greasy Pig Disease is a bacterial infection of the skin of the pig, which is known by a variety of other names - Greasy Skin, Exudative Epidermitis, Marmite Disease.

The primary cause of the disease is Staphylococcus hyicus, which is a common bacterium known to colonise the skin of many pigs without causing disease. As a group of diseases, it probably constitutes the most common skin ailment in pigs and is pathologically a dermatitis/epidermitis, the most significant feature of which is that it is non-irritant, making the skin condition easy to differentiate clinically from the other common pig skin disease - Sarcoptic Mange.

There are a wide variety of presentations of the disease, ranging from localised lesions on the body, including the ear tips, to whole body effects. The disease can occur in all age groups from soon after birth up to adults.

Development of Disease

The causative organism lives on the skin surface of the pig, but will require some form of trigger mechanism to produce disease. This may be:

1. Damage to the skin, either from fighting, other disease (e.g. Sarcoptic Mange, Pig Pox, Pityriasis Rosea) or injuries from floors, feeders or pen divisions.
2. A film of grease or faeces over the skin, under which the organism can multiply.
3. High humidity levels, producing moisture layer on the skin, within which the organism can proliferate.

Staph hyicus can be found on the skin of most pig populations but there appears to be various strains of the bacteria and it is possible that new variants can be introduced (usually with incoming stock) and cause outbreaks of disease.

Clinical Presentation

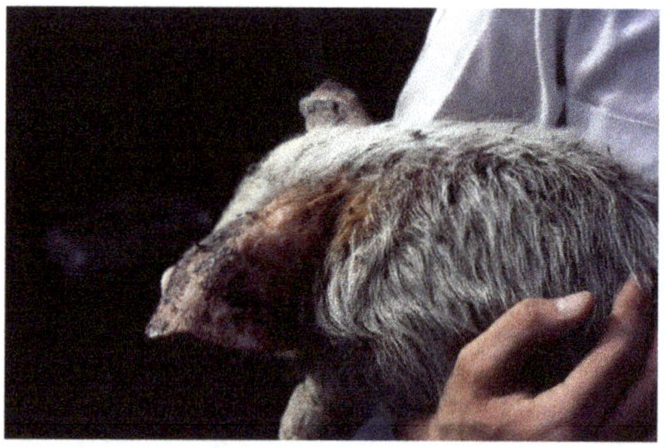

The classic form of Greasy Pig Disease in a weaner pig.

In the young pig, up to 7 weeks of age, the most common presentation of Greasy Pig Disease is one of a brown to black development of scab appearance, starting usually around the shoulder and neck and spreading to part, or the whole, of the body. The younger it occurs, the more serious the consequences. It can occur in the sucking piglet as young as 3-4 days and, at this age, can be lethal due to the disturbance that occurs to fluid balance.

At the young end of the scale, an acute ulcerative dermatitis can occur, particularly on the soft skin of the abdomen and chest and can easily be mistaken for a contact ulceration (e.g. due to disinfectant or lime washing). This form of the disease can be seen as young as 24 hours old and is often fatal.

Occasionally Staph hyicus can be involved in facial necrosis - the blackening of the skin of the face that result from teeth damage from litter mates where teeth clipping is not practiced. (Fusobacterium necrophorum is the more common cause of this disease).

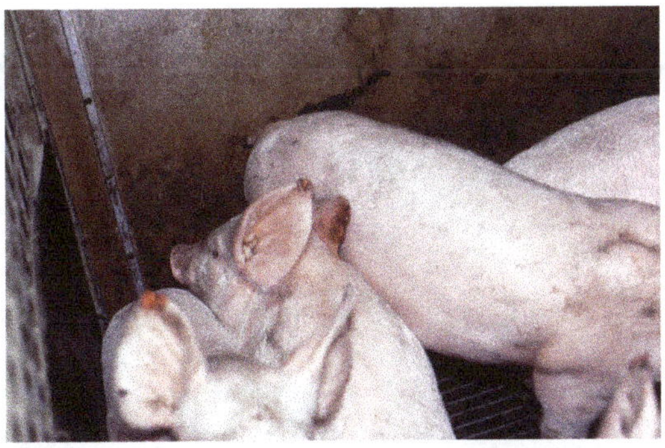

Erosion of the ear tips secondary to Staph Hyicus infection (ear tip necrosis).

Moreover, discreet dermatitis will commonly occur around the head, neck and shoulders of weaned pigs as a result of colonisation of skin wounds - themselves the result of fighting - by Staph hyicus that is present on the skin. This form of the disease will often appear to spread throughout a group of weaners.

As older weaners and growers, the aftermath of earlier widespread greasy pig disease in individuals can be seen as part of the healing process. The skin will appear shiny, often hairless and have a distinct orange tinge to it. This healing process can be extremely slow, to the extent that it can still be unhealed by the time the pig reaches slaughter weight, resulting in condemnation of the skin.

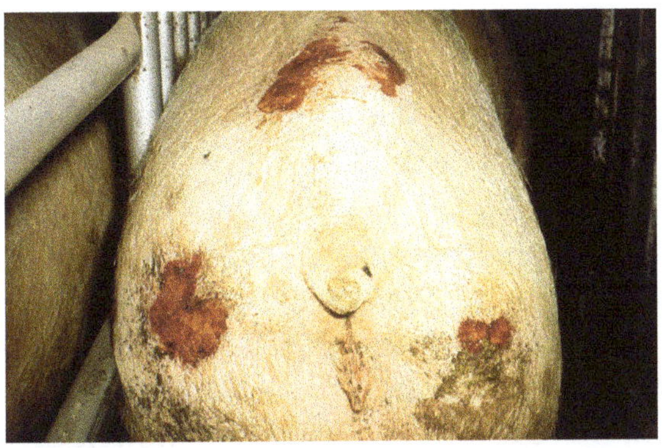

Severe ulcerative dermatitis in a sow associated with Staph hyicus infection.

Staph hyicus is implicated in the development of dry gangrene of the extremities - affecting the tail of young piglets, often before tail docking has occurred and the ear tips of weaners of 6-7 weeks upwards. The latter result is ear tip necrosis, an unsightly but apparently systemically harmless condition. Localised greasy pig lesions can also occur on the legs of weaned pigs, starting at the foot and gradually creeping upwards. Here, primary damage occurs close to the coronary band of the hoof, allowing penetration of skin colonising bacteria.

A further manifestation of greasy pig disease occurs in adults. Here, discreet superficial black lesions occur, usually over the back - are non-irritant and apparently harmless. They may reflect some form of immune incompetence in the individual animal - lesions often remain for life and prove intractable to treatment and litters of affected sows may well show classic greasy pig disease early in life. Rarely sows may present with acute ulcerative lesions similar to those seen in baby piglets but usually non-fatal.

Chronic recovering Greasy Pig Disease in a finisher likely to lead to skin condemnation and down-grading at slaughter.

Finally, a rare and fatal form of Staph hyicus infection is reported in individual sows, in which the whole of the skin is affected, becomes progressively thickened and wrinkled (like rhinoceros skin) and progressive severe loss of condition occurs. These animals are usually euthanased.

Treatment

As a bacterial infection, Greasy Pig Disease responds well to antibiotic treatment, provided:

- The antibiotic used is effective against the strain involved.
- Sufficient dose is given for sufficient duration.
- The antibiotic used penetrates into the skin in sufficient concentration.
- Treatment is given early.

Lincomycin, the penicillins and cephalosporins are most effective as injectable or orally applied treatments although the use of the latter - especially 3rd and 4th generation preparations must be reserved for use only as a last resort where alternative treatments have failed and laboratory testing supports use. Use of topical antibiotics direct onto the skin can be useful but there are no licensed products available in the UK for this purpose, although the advising veterinary surgeon may be able to prescribe "off licence" products.

For the young piglets, hydration is vital and any suckling pigs should be given supportive electrolytes if Greasy Pig Disease occurs. The use of soap (e.g. Savlon) can be a useful adjunct to treatment to remove the grease film over the skin and kill skin bacteria locally. Any concomitant disease that is acting to trigger Greasy Pig Disease (e.g. Sarcoptic Mange) must also be treated.

Prevention

If any one feature can be highlighted that will precipitate development of lesions, it is skin damage, either as a result of fighting, abrasion by floors or injury on pen divisions etc. Any control programmes must take this into account. Fighting can be reduced by minimising mixing and moving, maintaining stable groups and ensuring free access to feed, water and lying space. Teeth clipping or grinding will reduce damage to skin from fighting (it should be done within the first 3 days of life but only where the need is identified and supported by the veterinary surgeon - routine teeth reduction is not permissible). Fighting at weaning can be reduced by temporarily leaving pigs in the dark for an hour or two or spraying with a deodorant (e.g. Maskomal: Dupont), washing pigs in weak soap solution (e.g. Savlon) or disinfectant (e.g. 1% Virkon S) prior to mixing. The latter techniques can reduce skin colonisation as well.

In addition, in the face of a clinical problem, many farms will wash pigs in Savlon or Virkon S to reduce skin contamination and thus reduce the risk of secondary infection with Staphylococcus hyicus into fight wounds. Care is needed not to chill pigs and to avoid excessive humidity whilst maintaining satisfactory airflow.

Where lesions start around the feet, attention to flooring is essential. In farms where long standing intractable problems occurs, use of strategic (in feed or in water) medication may be appropriate in the short term in anticipation of disease developing and, in some cases, autogenous or farm specific vaccines can be prepared to apply either to sows or to young pigs. Correction of trigger factors however remains a priority.

Swine Dysentery

Swine dysentery (SD) is a severe, infectious disease characterized by mucohemorrhagic diarrhea and marked inflammation limited to the large intestine (cecum and/or colon). Spirochaetal colitis (SC) causes milder colitis in young -pigs.

Occurrence

Swine dysentery (SD) occurs only in swine although the etiologic agent infects and persists in rodents. All ages of swine may have SD although it seldom is apparent in piglets less than three weeks old. The disease occurs more frequently during the growing/finishing periods. Swine dysentery occurs worldwide in the major swine-raising countries and can cause very significant economic losses on affected herds. In the United States, SD now occurs only occasionally because of successful eradication and effective manure removal techniques. Outbreaks occur throughout the year, more often in late summer and early fall. With the increased interest in niche pork production (e.g. organic, antibiotic free) in the US, there has been an increase in incidence of SD in these systems.

Spirochaetal colitis associated with Brachyspira pilosicoli also has worldwide distribution, causing a milder, non-hemorrhagic colitis.

Etiology

Brachyspira (previously, Serpulina or Treponema) hyodysenteriae, a spirochete, is the etiologic agent. It has axial filaments, is Gram-negative, anaerobic and the most pathogenic strains are strongly beta-hemolytic. There are a few avirulent or weakly beta-hemolytic Brachyspira strains that are predominantly nonpathogenic. Other organisms in the large intestine, especially anaerobes, may facilitate colonization and lesion formation. Brachyspira hyodysenteriae has the ability to survive under a wide range of environmental conditions but is susceptible to heat, ultraviolet (UV) light, and desiccation, as well as soaps and disinfectants. Other spirochetes, especially B. innocens and B. pilosicoli are found in the large intestine of many swine and are easily mistaken for B. hyodysenteriae.

While considering a diagnosis of swine dysentery, one should be aware of recent work in describing a nonfatal, non-bloody, diarrheal disease in swine caused by Brachyspira (Serpulina) pilosicoli. This agent appears to be common in swine populations and is an occasional cause of catarrhal colitis in young growing swine. Virtually all statements related to B. hyodysenteriae are also true for B. pilosicoli with the exception that disease caused by the latter is much milder.

Epidemiology

Transmission of B. hyodysenteriae is by ingestion of infectious feces. It can persist in lagoon water for at least two months, moist feces for two months, and soil for 18 days. Mice have been reported to shed B. hyodysenteriae for 180 days and dogs for 13 days. The organism can be transmitted by birds, flies, and fomites. Carrier swine can transmit the agent for at least 90 days. Clinically normal, purchased carriers, including breeding stock, often are the source of initial exposure. Carrier sows often transmit to their piglets. Infected mice on premises may also be a source of infection.

Pathogenesis

Brachyspira hyodysenteriae is ingested with clinical signs developing 5-21 days later; incubation period is dependent upon dose but the infective dose is quite small. The organism reaches the large intestine where it colonizes, proliferates, penetrates the mucus layer and becomes closely associated with epithelial cells. Invasion may not be essential for lesion production. The exact mechanism of tissue destruction is not known but lipopolysaccharide in the organism probably is involved. Also, B. hyodysenteriae produces two known toxins and a hemolysin that may play roles. In typical lesions the organisms can be seen within epithelial cells and occasionally in the lamina propria. Lack of lesions at other sites suggests the entire pathogenesis is associated with the cecal and colonic lesions. B. pilosicoli appears to exert pathogenic effect by attachment to enterocytes and microerosion of epithelium.

Systemic effects presumably are the result of fluid and electrolyte imbalance caused by colitis. Fluid imbalance is the result of failure of the colon to absorb fluid, much of it from endogenous secretions, and explains the progressive dehydration and deaths that occur. Peracute deaths occasionally occur, perhaps caused by the toxins.

Clinical Signs

Diarrhea, usually with gray to yellow, mucoid feces often is the first sign noticed. With SD, diarrhea continues and quickly becomes mucohemorrhagic, with excess mucus and fresh blood apparent. In a small percent of the pigs, diarrhea may be preceded by tail twitching or a humped, gaunt appearance. Fresh, red blood in mucus-containing feces often is profuse and the perineal area may become blood stained. Signs that follow prolonged diarrhea are those associated with dehydration. These include sunken eyes, marked weakness, hollow flanks and weight loss. In advanced cases, appetite is erratic but the animals continue to drink. Advanced cases remain unthrifty, even with treatment.

Occasionally, sudden death is observed. In untreated herds morbidity is high and mortality can approach 50%. Spirochaetal colitis is generally seen as a mild persistent diarrhea with mucus. Affected pigs may be less thrifty but effects on growth performance may not be overtly obvious.

Lesions

Lesions are limited to large intestine, except for dehydration and the nonspecific reddening of the gastric mucosa. The mesentery and serosa are edematous and the serosa is rather opaque. Any or all parts of the large intestine (cecum, spiral colon, rectum) may be affected. It is heavy, thick walled, congested and edematous. The mucosa is thrown into folds and is covered, diffusely or in patches, with a layer of fibrin, necrotic debris and mucus. There often are flecks of blood in the mucus or on uncovered mucosa. The colon contains fibrinonecrotic debris and excessive mucus, often mixed with blood.

In acute cases the mucosa of the affected large intestine is reddened, edematous and swollen. There may be only small amounts of fibrin but excessive mucus. Blood usually can be seen in the feces. Pigs that die suddenly may be in surprisingly thrifty condition although lesions in the large intestine are extensive.

Microscopically, there is moderate nonsuppurative colitis and typhlitis, mucosal metaplasia, edema, and superficial epithelial necrosis. Spiral-shaped organisms can be demonstrated within crypts, enterocytes, and debris by silver stain.

Lesions of spirochaetal colitis are milder, with mild hyperemia and excess mucus observed on the mucosa of large intestine. Microscopically, there may be abundant spiral-shaped organisms colonizing epithelium, accompanied by mild nonsuppurative inflammation.

Diagnosis

A field diagnosis of SD often can be made on the basis of clinical signs and typical gross lesions. A simple aid in diagnosis is the demonstration of many spirochetes in stained smears made from colonic scrapings. Brachyspira hyodysenteriae is weakly Gram-negative so organisms are best demonstrated with crystal violet or Victoria blue 4-R stains. Other spirochetes are easily mistaken for B. hyodysenteriae, so it is advisable to get laboratory confirmation of the diagnosis, especially if the outbreak is an initial one or there is doubt about the diagnosis. B. hyodysenteriae often can be cultured and identified from rectal swabs or colonic scrapings from acutely affected, unmedicated pigs. Biochemical differentiation of pathogenic and nonpathogenic strains can be challenging. Polymerase chain reaction (PCR) techniques are available for detection of the organism and provide differentiation to the species level; their use is encouraged.

Several diarrheal diseases of swine must be differentiated from SD. These include proliferative enteritis, whipworm infestation, gastric ulcers, spirochaetal colitis (B. pilosicoli), and salmonellosis. Severe whipworm infections mimic mucohemorrhagic SD and worms are not grossly visible until three to four weeks after infestation. Salmonellosis lesions usually are not confined to the large intestine, as are swine dysentery lesions. Lesions of salmonellosis tend to extend deeper into the mucosa as ulcers and may be patchy in distribution with less mucus. Necropsy of several pigs and microscopic study of the colon often are helpful in separating the two diseases. Dual infections can occur. Spirochaetal colitis is best diagnosed by necropsy of acutely affected pigs, histopathology, isolation of the weakly beta-hemolytic causative organism, and definitive identification biochemically or by PCR.

Control

Preventing the introduction of B. hyodysenteriae to negative herds is a high priority. Additions to the herd should be only from herds known to be free of swine dysentery. They should be added only after a quarantine period of 30-60 days. Some veterinarians recommend treating high-risk animals with tiamulin or carbadox during quarantine to reduce the possibility of carriers being introduced.

Three methods are used to eliminate swine dysentery and have been quite successful in the US. Early weaned piglets (< three weeks) taken to a clean site and raised there usually remain free of B. hyodysenteriae. Meanwhile, the infected herd can be marketed, and the facilities cleaned and disinfected. This method allows one to retain valuable genetic stock. Segregated rearing with all in/all out production techniques has been quite successful in eliminating SD.

A second method relies on medication. Reduction of a herd to a minimal number of animals, followed by intensive treatment of those retained, is often used. Debilitated animals must be disposed of since they may not consume enough medicated feed or water. Prolonged, intensive treatment is expensive so numbers of swine to be treated should be minimized. Three widely used therapeutic agents include carbadox, lincomycin and tiamulin. Directions for their use should be followed carefully and every animal must be treated.

The third method of complete depopulation may be useful on premises with endemic swine dysentery where biosecurity and sanitation are difficult to implement. Depopulation should be done during warm, dry weather. During depopulation all facilities and equipment should be thoroughly cleaned and disinfected. The premises should be left vacant for at least two weeks, depending on the weather and level of achievable sanitation. Brachyspira hyodysenteriae usually will not survive for more than two weeks in soil or grass pens kept dry and relatively free of feces during dry, warm weather. Repopulation should be made with swine free of SD.

Any effort to eradicate SD requires an initial accurate diagnosis of SD, a commitment from the owners, and initial agreement on a protocol for accomplishing eradication and ongoing veterinary supervision. The eradication effort should be made during the warmest time of the year. A professionally-managed rodent control program is essential. Eradication must be coordinated with pig flow so that thorough cleaning and disinfection can be accomplished when the premises are vacant or the population minimal. If infected animals must be retained during the effort, a barrier system must be set up to separate clean from potentially contaminated areas.

Producers trying to control, but not eliminate infection with endemic swine dysentery can use preventive and therapeutic levels of antibiotics administered in feed and/or water. Those herds often can be treated intermittently to advantage. Although several products are FDA-approved for swine dysentery, carbadox, lincomycin, and tiamulin seem to be most effective. Antibiotic resistance is an increasing problem with this organism, making control programs (as opposed to elimination) a poor long term strategy. Vaccination has not proved to be reliably effective in control or elimination programs.

Caprine Arthritis and Encephalitis

Caprine arthritis and encephalitis (CAE) virus infection is manifested clinically as polysynovitis-arthritis in adult goats and less commonly as progressive paresis (leukoencephalomyelitis) in kids. Subclinical or clinical interstitial pneumonia, in durative mastitis ("hard udder"), and chronic wasting have also been attributed to infection with this virus. Most CAE virus infections, however, are subclinical. Infection with the CAE virus decreases the lifetime productivity of dairy goats and is a barrier to exportation of goats from North America.

CAE virus infection is widespread among dairy goats in most industrialized countries but rare among indigenous goat breeds of developing countries unless they have been in contact with imported goats. In countries such as Canada, Norway, Switzerland, France, and the USA, seroprevalence of CAE virus is >65%.

Etiology, Epidemiology and Pathogenesis

The CAE virus is an enveloped, single-stranded RNA lentivirus in the family Retroviridae. There are several, genetically distinct isolates of the virus that differ in virulence.

The CAE virus of goats is closely related to the ovine lentiviruses causing ovine progressive pneumonia and maedi-visna in North America and Europe, respectively. Cross-species transmission is possible through feeding of infected milk and colostrum. Therefore, the ovine and caprine lentiviruses are now commonly referred to as small ruminant lentiviruses.

CAE virus infection is widespread in dairy goat breeds but uncommon in meat- and fiber-producing goats. This has been attributed to genetics, management practices such as feeding colostrum and milk from a single dam to multiple kids, and industrialized farming practices (e.g., frequent introductions of new animals into a herd). Prevalence of infection increases with age but is not influenced by sex. Most goats are infected at an early age, remain virus positive for life, and develop disease months to years later.

The chief mode of spread of CAE is through ingestion of virus-infected goat colostrum or milk by kids. The feeding of pooled colostrum or milk to kids is a particularly risky practice, because a few infected does will spread the virus to a large number of kids. Horizontal transmission also contributes to disease spread within herds and may occur through direct contact, exposure to fomites at feed bunks and waterers, ingestion of contaminated milk in milking parlors, or serial use of needles or equipment contaminated with blood. Unlikely methods of transmission, as indicated by experimental studies, include in utero transmission to the fetus, infection of the kid during parturition, and infection through breeding or embryo transfer.

The pathogenesis of CAE is not fully understood. Virus-infected macrophages in colostrum and milk are absorbed intact through the gut mucosa. Infection is subsequently spread throughout the body via infected mononuclear cells. Periodic viral replication and macrophage maturation induces the characteristic lymphoproliferative lesions in target tissues such as the lungs, synovium, choroid plexus, and udder. Persistence of the CAE virus in the host is facilitated by its ability to become sequestered as provirus in host cells. Infection induces a strong humoral and cell-mediated immune response, but neither is protective.

Clinical Findings

Clinical signs are seen in ~20% of CAE virus–infected goats during their lifetime. The most common manifestation of infection is polysynovitis-arthritis, which is primarily seen in adult goats but can occur in kids as young as 6 moles. Signs of polysynovitis-arthritis include joint capsule distention and varying degrees of lameness. The carpal joints are most frequently involved. The onset of arthritis may be sudden or insidious, but the clinical course is always progressive. Affected goats lose condition and usually have poor hair coats. Encephalomyelitis is generally seen in kids 2–4 mo old but has been described in older kids and adult goats. Affected kids initially exhibit weakness, ataxia, and hindlimb placing deficits. Hypertonia and hyperreflexia are also common. Over time, signs progress to paraparesis or tetraparesis and paralysis. Depression, head tilt, circling, opisthotonos, torticollis, and paddling have also been described. The interstitial pneumonia component of CAE virus infection rarely produces clinical signs in kids. However, in adult goats with serologic evidence of CAE virus infection, chronic interstitial pneumonia that leads to progressive dyspnea has been documented. The "hard udder" syndrome attributed to CAE virus infection is characterized by a firm, swollen mammary gland and agalactia at the time of parturition. Milk quality is usually unaffected. Although the mammary gland may soften and produce close to normal amounts of milk, production remains low in many goats with indurative mastitis.

Lesions

Pathologic lesions of CAE virus infection are generally described as lymphoproliferative with degenerative mononuclear cell infiltration. Lesions in joints are characterized by thickening of the joint capsule and marked proliferation of synovial villi. In chronic cases, soft-tissue calcification involving joint capsules, tendon sheaths, and bursae is not uncommon. Severe cartilage destruction, rupture of ligaments and tendons, and periarticular osteophyte formation has also been described in advanced cases. Microscopic features of articular lesions include synovial cell hyperplasia, subsynovial mononuclear cell infiltration, villous hypertrophy, synovial edema, and synovial necrosis. Gross lesions associated with the neurologic form of CAE include asymmetric, brownish pink, swollen areas, most commonly in the cervical and lumbosacral spinal cord segments. Histopathologically, these lesions are characterized by multifocal, mononuclear cell inflammatory infiltrates and varying degrees of demyelination. On gross examination, lungs of affected goats are firm and gray-pink with multiple, small, white foci, and do not collapse. The bronchial lymph nodes are invariably enlarged. Histologic findings include chronic interstitial pneumonia with mononuclear cell infiltration in alveolar septae and in perivascular and peribronchial regions. In does with udder induration, mononuclear infiltration of periductular stroma obliterates normal mammary tissue.

Diagnosis

A presumptive diagnosis can be based on clinical signs and history. Infectious arthritis caused by Mycoplasma spp and traumatic arthritis are differential diagnoses for arthritis induced by CAE virus. Differential diagnoses for the progressive paresis and paralysis exhibited by young kids should include enzootic ataxia, spinal cord abscess, cerebrospinal nematodiasis, spinal cord trauma, and congenital anomalies of the spinal cord and vertebral column. If the neurologic examination indicates brain involvement, polioencephalomalacia, listeriosis, and rabies should be considered as possible causes. The pulmonary form of caseous lymphadenitis may have a similar clinical presentation to the pulmonary form of CAE in adult goats.

Both an agar gel immunodiffusion test and ELISA for CAE virus are considered sufficiently reliable for use in control programs. The agar gel immunodiffusion test is reported to be more specific but less sensitive than the ELISA. A positive test result in an adult goat implies infection but does not confirm that the clinical signs are caused by CAE virus. Kids infected at birth develop a measurable antibody response 4–10 wk after infection. However, positive test results in kids <90 days old usually reflect colostral antibody transfer. Negative test results do not reliably exclude CAE virus infection, because the time for postinfection seroconversion is variable and occasional goats have a very low titer that may not be detectable. Low antibody titers are common in late pregnancy. Because of the limitations of serologic testing, definitive diagnosis of clinical CAE requires demonstration of characteristic lesions in biopsy specimens or at necropsy. Virus isolation or PCR to demonstrate presence of viral antigen in tissues may be used to further substantiate the diagnosis.

Treatment and Control

There are no specific treatments for any of the clinical syndromes associated with CAE virus infection. However, supportive treatments may benefit individual goats. The condition of goats with the polysynovitis-arthritis may be improved with regular foot trimming, use of additional bedding, and administration of NSAIDs such as phenylbutazone or aspirin. Goats with encephalomyelitis can be maintained for weeks with good nursing care. Antimicrobial therapy is indicated to treat secondary bacterial infections that may complicate the interstitial pneumonia or indurative mastitis components of CAE virus infection. Providing high-quality, readily digestible feed to goats positive for CAE virus may delay the onset of the wasting syndrome.

In commercial herds, one or more of the following have been recommended for control of CAE: 1) permanent isolation of kids beginning at birth; 2) feeding of heat-treated colostrum (45°C [113°F] for 60 min) and pasteurized milk; 3) frequent serologic testing of the herd (semiannually), with identification and segregation of seronegative and seropositive goats; and 4) eventual culling of seropositive goats. If the control program includes segregation of herds into seropositive and seronegative groups, groups should be separated by a minimum of 6 ft (1.8 m), and shared equipment should be disinfected using phenolic or quaternary ammonium compounds.

Caseous Lymphadenitis

Caseous lymphadenitis (CL) is a chronic, contagious bacterial disease that manifests clinically as abscesses of peripheral and/or internal lymph nodes and organs. The characteristic purulent material is very thick and nonodorous. Whereas the peripheral form presents as abscesses of single or multiple peripheral palpable lymph nodes, internal CL typically manifests as chronic weight loss and ill thrift. Culture of active lesions for Corynebacterium pseudotuberculosis is diagnostically definitive. When eliminating animals from the herd/flock is undesirable, treatment consists of consistent, sustained antimicrobial therapy to reduce the numbers of active draining lesions and isolation from other herd mates until lesions are dry and resolved.

Caseous lymphadenitis (CL) is a chronic, contagious disease caused by the bacterium Corynebacterium pseudotuberculosis. Although prevalence of CL varies by region and country, it is found worldwide and is of major concern for small ruminant producers in North America. The disease is characterized by abscess formation in or near major peripheral lymph nodes (external form) or within internal organs and lymph nodes (internal form). Although both the external and internal forms of CL occur in sheep and goats, the external form is more common in goats, and the internal form is more common in sheep.

Economic losses from CL include death, condemnation and trim of infected carcasses, hide and wool loss, loss of sales for breeding animals, and premature culling of affected animals from the herd or flock. Once established on a farm or region (endemic), it is primarily maintained by contamination of the environment with active draining lesions, animals with the internal form of the disease that contaminate the environment through nasal discharge or coughing, the ability of the bacteria to survive harsh environmental conditions, and lack of strict biosecurity necessary to reduce the number and prevent introduction of new cases. Although CL is typically considered a disease of sheep and goats, it also occurs more sporadically in horses, cattle, camelids, swine, wild ruminants, fowl, and people. Because of its zoonotic potential, care should be taken when handling infected animals or purulent exudate from active, draining lesions.

Etiology and Pathogenesis of Caseous Lymphadenitis of Sheep and Goats

C pseudotuberculosis is a gram-positive, facultative, intracellular coccobacillus. Two biotypes have been identified based on the ability of the bacteria to reduce nitrate: a nitrate-negative group that infects sheep and goats, and a nitrate-positive group that infects horses. Isolates from cattle are a heterogeneous group. All strains produce an exotoxin called phospholipase D that enhances dissemination of the bacteria by damaging endothelial cells and increasing vascular permeability. The bacterium has a second virulence factor, which is an external lipid coat that provides protection from hydrolytic enzymes in host phagocytes. Replication of bacteria occurs in the phagocytes, which then rupture and release bacteria. The ongoing process of bacterial replication, followed by attraction and subsequent death of inflammatory cells, forms the characteristic abscesses associated with CL.

To establish infection, C pseudotuberculosis must penetrate skin or mucous membranes. The most common site of entry is the skin after an injury that may result from shearing, tagging, tail docking, castration, or other environmental hazards resulting in skin trauma. Contact with purulent material draining from open, active lesions most commonly serves as the source of bacteria through these breaches in the skin. Although less common, entry across mucous membranes from inhalation or ingestion of the bacteria also serves as a means of infection.

Once the bacteria have entered the body, they move to the lymph nodes via the regional draining lymphatic system. Internally, the bacteria establish infection not only in the lymph nodes but also in the viscera. The incubation period varies from 1 to 3 months, culminating in development of encapsulated abscesses. C pseudotuberculosis is hardy in the environment and can survive on fomites such as bedding and wood for 2 months and in soil for 8 months. The presence of organic material, shade, and moisture favor and enhance survival.

Clinical Findings of Caseous Lymphadenitis of Sheep and Goats

The hallmark clinical finding in cases of external caseous lymphadenitis is the development of abscesses in the region of peripheral lymph nodes. Common sites of development include the submandibular, parotid, prescapular, and prefemoral nodes. Less commonly, abscessation of supramammary or inguinal lymph nodes occurs, in addition to an occasional ectopic location along the lymphatic chain. If left untreated, these lesions eventually mature into open draining abscesses. The purulent material from these lesions has no odor and varies in consistency from soft and pasty (more common in goats) to thick and caseous (more common in sheep). Once natural draining occurs, the skin lesion heals with scarring.

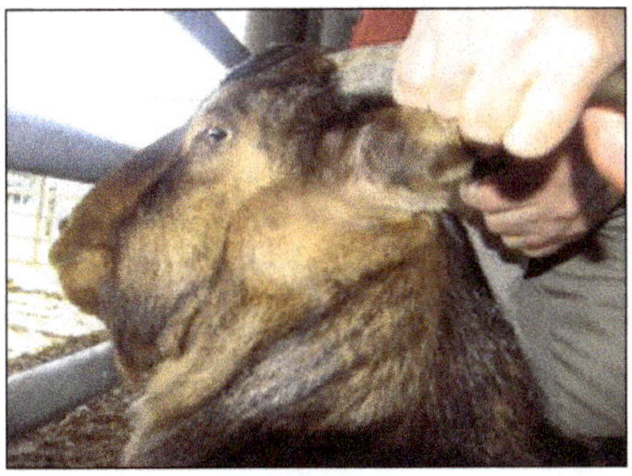
Caseous lymphadenitis lesions, external form, parotid lymph node.

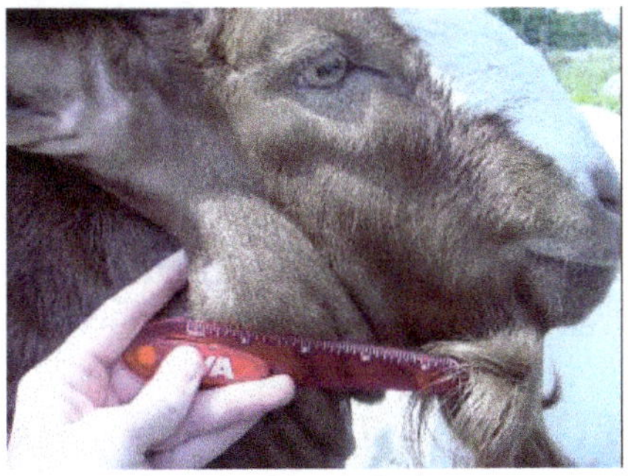
Caseous lymphadenitis lesions, external form, submandibular lymph node.

Recurrence is common, which can be months later. CL should be highly suspected in a sheep or goat with abscessation in these regions. Although other bacteria may cause abscessation in these locations (and in other animals), because of the ramifications of the presence of this disease within a herd or flock, these cases should be handled as CL until proved otherwise.

The internal form of CL most commonly presents as chronic weight loss and failure to thrive. The presence of other clinical signs depends on the organs of involvement, which may include any of the major organ systems. Lung abscessation is a common site of visceral involvement in internal CL; therefore, signs of chronic ill thrift with cough, purulent nasal discharge, fever, and tachypnea with increased lung sounds may be noted. The internal form is more common in sheep and has been termed the "thin ewe syndrome."

The incidence of abscesses and development of clinical disease with either the external or internal form increases with age.

Lesions

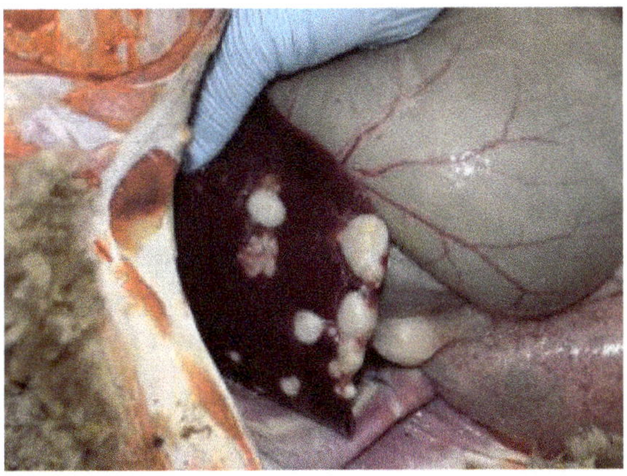
Caseous lymphadenitis lesions, internal form, liver of a sheep.

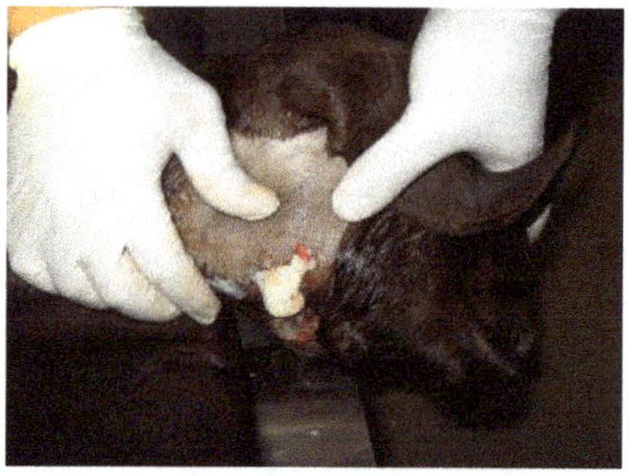

Draining Caseous Lymphadenitis Abscess.

In sheep, abscesses often have the classically described laminated "onion-ring" appearance in cross section, with concentric fibrous layers separated by inspissated caseous exudate. In goats, the abscesses are less organized, and the exudate may be soft and paste-like.

Diagnosis of Caseous Lymphadenitis of Sheep and Goats

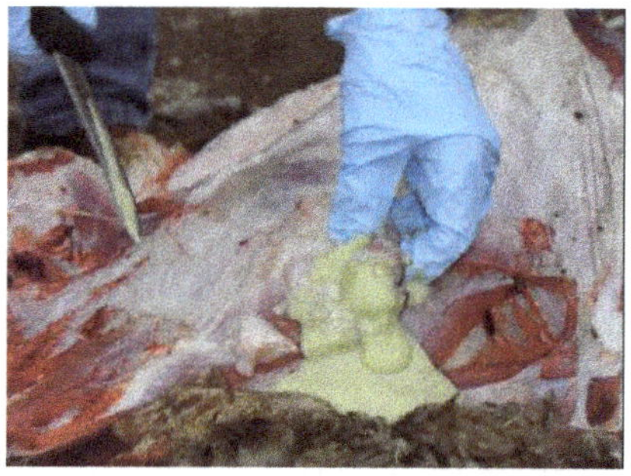

Caseous lymphadenitis abscess of the prefemoral lymph node, goat.

- Physical examination of lesions associated with lymph nodes;
- Bacterial culture of suspected lesions;
- Serologic testing.

The presence of an external abscess on a small ruminant is highly suggestive of caseous lymphadenitis, especially in locations of peripheral lymph nodes. However, definitive diagnosis is only by bacteriologic culture of purulent material from an intact abscess. Although other pyogenic organisms such as Trueperella pyogenes (formerly Arcanobacterium pyogenes), Staphylococcus aureus, Pasteurella multocida, and anaerobes such as Fusobacterium necrophorum can cause abscessation, affected animals should be kept isolated pending culture results.

Animals with visceral abscesses pose a greater diagnostic challenge. Radiography and ultrasonography can be useful to detect internal lesions. Culture of a transtracheal aspirate obtained from an animal with pneumonia can help determine whether CL is the cause. Excluding other causes of chronic weight loss and ill thrift in the face of proper nutrition and good appetite such as Johne's disease, parasitism, and poor dentition further raise suspicion.

In the absence of accessible abscesses for bacterial culture, definitive diagnosis of active cases of caseous lymphadenitis is challenging. Although many diagnostic tools are available, results of these tests must be interpreted with caution and with consideration of herd or flock history, the presence or absence of active infection within the herd or flock, and vaccination status. A synergistic hemolysin inhibition (SHI) test that detects antibodies to the phospholipase D exotoxin is available at many diagnostic laboratories. Positive titers indicate past resolved infections, recent exposure, recent vaccination, or active lesions or their development. Titers of 1:256 or higher have been correlated in past studies with the presence of active, developing abscesses; however, in a more recent study, a high titer was poorly correlated with presence or development of abscesses over an 18-month period.

When the status of an animal with a positive titer is in doubt, the titer should be repeated in 2–4 weeks. If the titer is rising and clinical signs of abscesses are noted, then CL can be assumed to be the cause. False-negative results can occur if testing is done in the first 2 weeks after exposure before the animal has seroconverted. Also, animals with chronic, walled-off abscesses can have a false-negative result. Colostrum titers usually disappear by 3–6 months of age, so serologic testing of lambs or kids <6 months old should be interpreted with caution.

Treatment and Control of Caseous Lymphadenitis of Sheep and Goats

Treatment

- Antimicrobial treatment (intralesional and/or systemic);
- Supportive care.

Control

- Strict biosecurity measures;
- Elimination of diseased animals from the herd/flock;
- Vaccination;
- Disinfection of shearing equipment and other instruments used for production procedures (castration, ear tagging, etc) between animals;
- Removal of hazards in the environment that could potentially injure the skin;
- Prepurchase examination for lesions, serologic screening and a period of quarantine before introduction of new animals.

Once a diagnosis of caseous lymphadenitis has been established, owner education stressing the persistent, recurrent nature of the disease is necessary. The most practical approach for commercial animals infected with CL is to cull them from the herd or flock. However, animals with draining abscesses should not be sent through sale barns until draining has ceased and the wound has healed. Treatment of individual animals should be undertaken with the understanding that CL is not considered a "curable" disease. Animals with genetic or emotional value are treated mainly for aesthetic reasons and to limit their infectivity to the rest of the herd or flock. Treatment options have included lancing and draining, surgical excision, formalin injection of lesions, systemic antibiotics, and intraregional antibiotics.

If external abscesses are lanced and drained, the cavity should be lavaged with dilute iodine solution and the animal isolated in an area that can be disinfected until the lesion stops draining and heals. Drained purulent material should be carefully collected and disposed of. Dilute bleach and chlorhexidine solutions are effective disinfectants of hard surfaces and fomites, but the presence of organic material on these surfaces inactivates them and drastically reduces or prohibits effectiveness. Intact accessible abscesses can be surgically removed; however, this option is more expensive, and undetected abscesses are often present and continue to develop. Recurrence rates with either lancing or surgical removal are high.

The practice of injecting abscesses with formalin should be strongly discouraged, because the FDA has zero tolerance for extra-label use of a potent carcinogen in food-producing animals. The efficacy of systemic antimicrobial therapy and, more recently, intralesional antimicrobial therapy has been investigated. It must be noted here that the use of any antimicrobial for treatment of CL is extra-label; therefore, strict adherence to published guidelines on withdrawal times and an established veterinarian-client-patient relationship are mandatory. Long-term administration of procaine penicillin G and rifampin has been successful in some cases. Penicillin alone, although effective in vitro, is unlikely to penetrate the capsule of developed abscesses, as are many, if not most, of the water-soluble or moderately lipid-soluble antimicrobials.

Recent studies have shown that administration of one dose of tulathromycin at 2.5 mg/kg, either SC directly into the abscess cavity, or two doses at 2.5 mg/kg, administered at the same time, one SC or one intralesionally can resolve the lesions without lancing the abscess. Further, effective concentrations of tulathromycin can be achieved within walled-off abscesses caused by C pseudotuberculosis after a single dose at 2.5 mg/kg, SC. The highly lipid-soluble property of tulathromycin may be particularly helpful in cases of internal CL, when abscesses are not accessible for other forms of treatment. Despite the efficacy of intralesional and parenteral administration of tulathromycin in many cases, recurrence remains a problem. Therefore, use of these drugs cannot be considered curative but rather an acceptable alternative to manage cases of CL when culling from the herd or flock is not an acceptable option for the owner.

Because of the nature of the causative organism, common means of exposure, chronicity of the disease, and difficulty in completely eliminating the organism from individual animals, control of CL focuses on strict biosecurity measures. The overriding goals of any control program are to eliminate the disease from the herd or flock and to reduce the number of new cases either from the spread of disease or introduction to the farm.

Ideally, animals identified as infected should immediately be culled. If immediate removal is not possible, infected animals should be isolated from the rest of the herd or flock. Diligence in this practice will eventually result in decreased prevalence as animals that develop active cases are identified and removed and given there are no new animals incubating the disease introduced to the premises.

When elimination through culling is not a viable option for the owner, control of CL is challenging at best. Dividing the herds or flocks into “clean” and “infected” groups and eliminating older and less genetically valuable animals over time is one control strategy.

Lambs and kids from infected dams can be raised on pasteurized colostrum and milk away from infected animals. However, the internal form of CL and animals incubating the disease can maintain infection within the “asymptomatic clean” group and limit the success of this approach.

Commercial CL vaccines are currently licensed for use in sheep and goats. These vaccines should only be used in the species they are labeled for, because adverse reactions have been reported in goats given vaccine labeled for sheep. Rigidly adhering to vaccination schedules according to the manufacturer's labeling can help reduce the prevalence and incidence of CL within herds or flocks. However, it is important to emphasize that efficacy of these vaccines is not 100%, and vaccination will not clear infected animals. Vaccination of young replacement stock should be considered, and older infected animals should be gradually culled as economics allow. Once the disease is at a low prevalence rate, vaccination should be stopped and all seropositive unvaccinated animals culled. In "clean" herds or flocks that have no history of CL, vaccination is not recommended.

The risks of disease transmission among animals should be recognized when shearing or dipping, and management practices should be adjusted accordingly. Animals with noted lesions should be shorn last, and clipper blades disinfected between animals. Shearers should recognize the hazards associated with contact with purulent material and the possibility of acting as mechanical vectors, either on clothing or via equipment, for spread of the bacteria to new animals. Further, dipping tank solutions should be kept as fresh as possible, because C pseudotuberculosis can survive within them and serve as a source of infection of freshly shorn sheep that have skin abrasions.

Owners should remove hazardous items (barbed wire, exposed nails, rough feeders) from the environment to decrease injury and potential CL transmission from the presence of bacteria on these fomites.

One of the most common ways CL can be introduced into a previously "clean" herd or flock, or reintroduced to one in which CL has been reduced or eliminated, is through the addition of replacement stock. Often, animals from other farms that are asymptomatic on arrival are incubating the disease and then manifest infection weeks to months later. Purchasing animals from sources with unknown histories is hazardous to maintaining a "clean" herd or flock. Newly arrived animals should be examined thoroughly for signs of CL, such as abscesses or scars near peripheral lymph nodes. They should remain isolated from the rest of the herd or flock until their serologic status is determined and only animals that are seronegative with no evidence of present or past CL lesions should be allowed to enter the herd or flock.

Equine Arthritis

Arthritis is a degenerative joint disease involving the breakdown of soft cartilage between the bones; the disorder can cause pain and lead to lameness in horses. Arthritis in horses also goes by the names DJD, equine arthritis, joint disease, osteoarthritis, OA, ringbone, and bone spavin.

Symptoms of Arthritis in Horses

In its early stages, arthritis may appear as a slight stiffness that the horse works out of once it is warmed up. In the joint, the cartilage material is breaking down but not repairing itself efficiently. Eventually, as the cartilage becomes more damaged, the bone beneath the pad of cartilage becomes damaged as well. The horse will then experience more discomfort and there may be heat in the joint as inflammation occurs. In more advanced cases, small bony growths called osteophytes may be seen under X-ray and may be felt around the affected joint. As the disease progresses, the horse may experience more discomfort and lameness.

Causes of Arthritis

Arthritis is caused by the slow wearing of cartilage the tough, flexible tissue that is the shock-absorbing and sliding surface between the bones that meet at a joint. Over time, compression and stress wear away the protective cartilage. Arthritis most commonly occurs in the weight-bearing joints of the legs and in the hooves. Extra stress or injury to any joint can lead to arthritis. Pain occurs as the cartilage thins, resulting in pain, inflammation, stiffness, and eventually, lameness.

Horses that are used for jumping, dressage, racing, reining, and other high-impact sports are more susceptible to arthritis in all the weight-bearing joints, especially fetlocks, knees, and hocks. Equine arthritis can affect any joint in the body, including the knees, shoulder joints, neck, fetlocks, hocks, back, and stifles. The fetlocks, knees, coffin (within the hoof), and hocks are most commonly affected. These are all weight-bearing joints.

Treatment

The goal of any treatment is to reduce the inflammation in joints, relieve pain, and slow the cartilage breakdown. There is no cure for arthritis at this time. If you notice that your horse is having any discomfort (even discomfort that wears off quickly), it's a good idea to consult your vet. You may be able to identify arthritis early and slow its progress with drugs that reduce inflammation.

Once the cartilage in a joint is gone, it is difficult to repair it. Further damage and discomfort may be waylaid by injections of healing or lubricating fluids such as hyaluronan, glucosamine, and corticosteroids, to name the most commonly used therapies. For very short-term pain relief, quipalazone (bute or Butazone) may be used, but long-term use can lead to kidney damage and cannot be used when competing. For longer-term treatment, new options include firocoxib (an NSAID) and diclofenac sodium, available in a topical cream. In addition:

- Stem cell therapy is also possible, although a more costly alternative.
- While cartilage transplant and other therapies are now possible, they are costly and the outcome is not guaranteed. Because arthritis is a common problem, research into its exact causes and more effective treatments is ongoing.
- Some people have found herbal preparations to be effective for relieving discomfort, but it may not be wise to use them without consulting a veterinarian.

In addition to veterinary treatment, a reduction in workload may be necessary, especially if the work includes jumping, traveling over hard terrain, or other activity that over-stresses the joints. This may mean the end of a horse's competitive career, although light exercise is important to maintain joint flexibility. At times when the horse may be lame, it should not be ridden. Keep the horse on a soft footing, with extra, but not excessive bedding that may be hard to walk through. Proper trims and shoeing by a knowledgeable farrier may also be helpful.

Younger horses can become arthritic with something called septic arthritis, which is difficult to treat. Antibiotics must be injected into the joint capsule to combat bacterial infection. Irreversible damage may be done to the joints of a foal once septic arthritis has occurred.

Prevention

Arthritis, to some degree, is almost unavoidable in older horses. Horses with conformation faults may put extra stress on joints, yet another reason why breeders strive for foals with good conformation. Proper hoof trims and shoeing, good conformation, providing good footing, while also providing good basic maintenance, can also help to delay the onset of arthritis.

Marek's Disease

Marek's Disease Virus (MDV) is a highly contagious viral infection that predominantly affects chickens but can also affect pheasants, quail, game fowl and turkeys. It is one of the most common diseases that affect poultry flocks worldwide. Clinical disease is not always apparent in infected flocks, however subclinical disease is often more economically important as it reduces weight gain and egg production. Mortality rates can be very high in susceptible birds. Marek's Disease (MD) results in enlarged nerves and in tumour formation in nerve, organ, muscle and epithelial (cells that line the internal and external surfaces of the body) tissue. Clinical signs include paralysis of the legs, wings and neck, weight loss, grey iris or irregular pupil, vision impairment and the skin around feather follicles can be raised and roughened. Affected birds are immunosuppressed and as a consequence are more susceptible to other infectious diseases.

Etiology

Marek's disease virus is a member of the genus Mardivirus within the subfamily Alphaherpesvirinae. Within the genus Mardivirus are three closely related species previously designated as three serotypes of Marek's disease virus. Gallid herpesvirus 2 (MDV-1) represents all virulent Marek's disease virus strains and is further divided into pathotypes, designated as mild (m), virulent (v), very virulent (vv), and very virulent plus (v+). Gallid herpesvirus 3 (MDV-2) and Meleagrid herpesvirus 1 (turkey herpesvirus, MDV-3) represent avirulent virus strains isolated from chickens and turkeys, respectively, and are commonly used as vaccines against Marek's disease.

Transmission and Epidemiology

The disease is highly contagious and readily transmitted among chickens. The virus matures into a fully infective, enveloped form in the epithelium of the feather follicle, from which it is released into the environment. It may survive for months in poultry house litter or dust. Dust or dander from infected chickens is particularly effective in transmission. Once the virus is introduced into a chicken flock, regardless of vaccination status, infection spreads quickly from bird to bird. Infected chickens continue to be carriers for long periods and act as sources of infectious virus. Shedding of infectious virus can be reduced, but not prevented, by prior vaccination. Unlike virulent strains of Marek's disease virus, which are highly contagious, turkey herpesvirus is not readily transmissible among chickens (although it is easily transmitted among turkeys, its natural host). Attenuated Marek's disease virus strains vary greatly in their transmissibility among chickens; the most highly attenuated are not transmitted. Marek's disease virus is not vertically transmitted.

Pathogenesis

Currently, four phases of infection in vivo are recognized: 1) early productive-restrictive virus infection causing primarily degenerative changes, 2) latent infection, 3) a second phase of cytolytic, productive-restrictive infection coincident with permanent immunosuppression, and 4) a proliferative phase involving nonproductively infected lymphoid cells that may or may not progress to the point of lymphoma formation. Productive infection may occur transiently in B lymphocytes within a few days after infection with virulent Marek's disease virus strains and is characterized by antigen production, which leads to cell death. Because few if any virions are produced, this has also been termed a restrictive-productive infection. Productive infection also occurs in the feather follicle epithelium, in which enveloped virions are produced. Latent infection of activated T cells is responsible for the longterm carrier state. No antigens are expressed, but virus can be recovered from such lymphocytes by co-cultivation with susceptible cells in tissue cultures. Some T cells, latently infected with oncogenic Marek's disease virus strains, undergo neoplastic transformation. These transformed cells, provided they escape the immune system of the host, may multiply to form characteristic lymphoid neoplasms. Cell-mediated and humoral immune responses are both directed against viral antigens, with cell-mediated immunity probably being the most important.

Clinical Findings

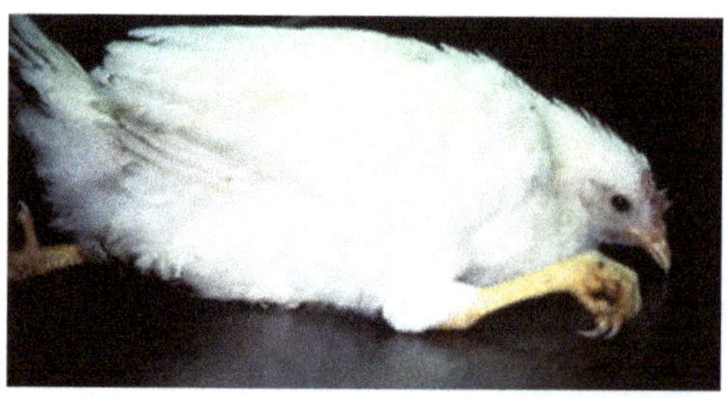

Marek's disease, leg paresis, chicken.

The incidence of Marek's disease is quite variable in commercial flocks and depends on strain and dose of virus, age at exposure, maternal antibody, host gender and genetics, strain and dose of vaccine virus, and several environmental factors, including stress. In addition to lymphoid neoplasms, Marek's disease virus can also induce other clinically distinct disease syndromes, including transient paralysis, early mortality syndrome, cytolytic infection, atherosclerosis, and persistent neurologic disease. Typically, affected birds show only depression before death, but a transient paralysis syndrome has been associated with Marek's disease; chickens become ataxic for periods of several days and then recover. This syndrome is rare in immunized birds. Death is usually the result of paralysis, rendering the birds unable to reach food and water.

Lesions

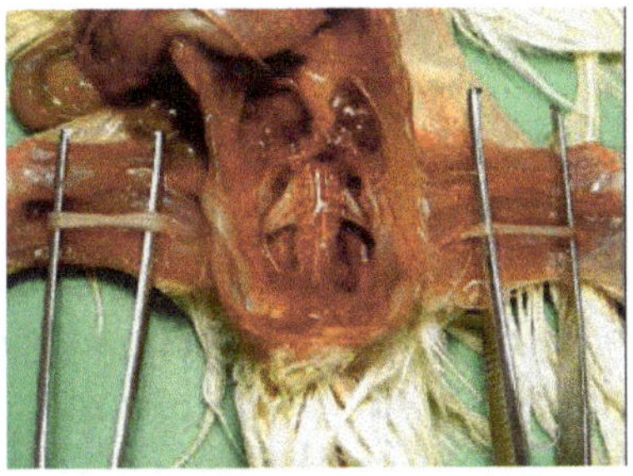

Marek's disease, peripheral nerve enlargement, chicken.

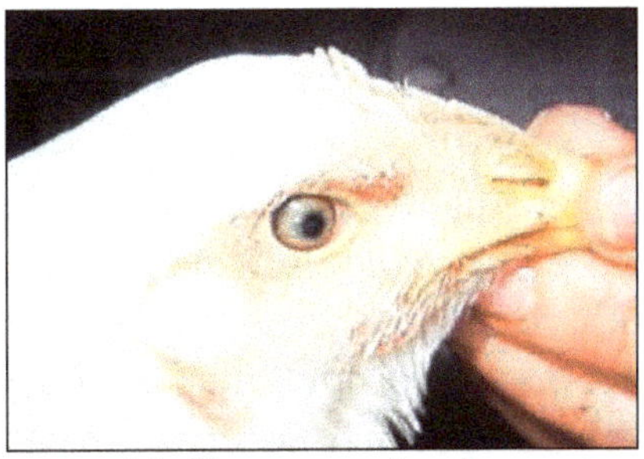

Marek's disease, eye involvement, chicken.

Enlarged nerves are one of the most consistent gross lesions in affected birds. Various peripheral nerves, but particularly the vagus, brachial, and sciatic, become enlarged and lose their striations. Diffuse or nodular lymphoid tumors may be seen in various organs, particularly the liver, spleen, gonads, heart, lung, kidney, muscle, and proventriculus. Enlarged feather follicles (commonly termed skin leukosis) may be noted in broilers after defeathering during processing and are a cause for condemnation. The bursa is only rarely tumorous and more frequently is atrophic. Histologically, the lesions consist of a mixed population of small, medium, and large lymphoid cells plus plasma cells and large anaplastic lymphoblasts. These cell populations undoubtedly include tumor cells and reactive inflammatory cells. When the bursa is involved, the tumor cells typically appear in interfollicular areas.

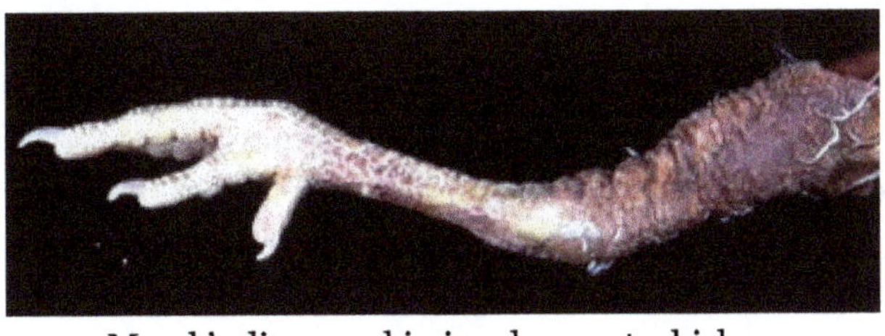

Marek's disease, skin involvement, chicken.

Diagnosis

For the diagnosis of Marek's disease, it is critical to diagnose the tumors and not the infection because Marek's disease is considered ubiquitous within commercial poultry flocks. Usually, diagnosis is based on enlarged nerves and lymphoid tumors in various viscera. The absence of bursal tumors helps distinguish this disease from lymphoid leukosis, although the presence of bursal tumors does not exclude Marek's disease. Marek's disease can develop in chickens as young as 3 wk old, whereas lymphoid leukosis typically is seen in chickens >14 wk old. Reticuloendotheliosis, although rare, can easily be confused with Marek's disease, because both diseases feature enlarged nerves and T-cell lymphomas in visceral organs. A diagnosis based on typical gross lesions may be confirmed histologically or preferably by demonstration of predominant T-cell populations and Marek's viral DNA in lymphomas by histochemistry and PCR, respectively. There is a quantitative association between viral load and Marek's disease tumors; most tumor-bearing chickens have high viremia titers and are usually PCR positive. Thus, the demonstration of high quantities of virus, viral DNA, or viral antigens in tumor cells and the exclusion of other relevant tumor viruses should be sufficient for a specific diagnosis of Marek's disease. Furthermore, Marek's disease lymphomas usually lack evidence of clonally integrated avian retroviruses or alteration of the cellular oncogene c-myc.

Control

Vaccination is the central strategy for the prevention and control of Marek's disease. The efficacy of vaccines can be improved, however, by strict sanitation to reduce or delay exposure and by breeding for genetic resistance. Probably the most widely used vaccine consists of turkey herpesvirus (HVT), which has seen rapidly increased use as a backbone in recombinant vaccines featuring the insertion of genes from other poultry viruses, such as Newcastle disease virus, infectious bursal disease virus, or infectious laryngotracheitis virus. These recombinant vaccines offer protection against both Marek's disease virus and the inserted virus. Bivalent vaccines consisting of HVT and either the SB-1 or 301B/1 strains of Gallid herpesvirus 3 have been used to provide additional protection against challenge with virulent Marek's disease virus isolates. The most protective commercial vaccine currently available appears to be CVI988/Rispens, an attenuated Marek's disease virus strain that is also commonly mixed with HVT at vaccination. Because vaccines are administered at hatching and require 1–2 wk to produce an effective immunity, exposure of chickens to virus should be minimized during the first few days after hatching.

Vaccines are also effective when administered to embryos at the 18th day of incubation. In ovo vaccination is now performed by automated technology and is widely used for vaccination of commercial broiler chickens, mainly because of reduced labor costs and greater precision of vaccine administration.

Proper handling of vaccine during thawing and reconstitution is crucial to ensure that adequate doses are administered. Cell-associated vaccines are generally more effective than cell-free vaccines, because they are neutralized less by maternal antibodies. Under typical conditions, vaccine efficacy is usually >90%. Since the advent of vaccination, losses from Marek's disease have been reduced dramatically in broiler and layer flocks. However, disease may become a serious problem in individual flocks or in selected geographic areas (eg, the Delmarva broiler industry). Of the many causes proposed for these excessive losses, early exposure to very virulent virus strains appears to be among the most important.

References

- Animal-disease, science: britannica.com, Retrieved 1 August, 2019
- Anthrax: beefresearch.ca, Retrieved 22 May, 2019

- Bluetongue: iastate.edu, Retrieved 18 January, 2019
- Ovine-brucellosis, livestock-biosecurity: agric.wa.gov.au, Retrieved 8 August, 2019
- Greasy-pig-disease, pigs: nadis.org.uk, Retrieved 17 June, 2019
- Swine-dysentery: iastate.edu, Retrieved 23 February, 2019
- Caprine-arthritis-and-encephalitis: msdvetmanual.com, Retrieved 3 March, 2019
- Caseous-lymphadenitis-of-sheep-and-goats, lymphadenitis-and-lymphangitis: msdvetmanual.com, Retrieved 24 March, 2019
- Equine-arthritis: thesprucepets.com, Retrieved 5 July, 2019
- Mareks-disease-virus-or-mdv, types-of-disease, disease, health: poultryhub.org, Retrieved 11 April, 2019

Permissions

All chapters in this book are published with permission under the Creative Commons Attribution Share Alike License or equivalent. Every chapter published in this book has been scrutinized by our experts. Their significance has been extensively debated. The topics covered herein carry significant information for a comprehensive understanding. They may even be implemented as practical applications or may be referred to as a beginning point for further studies.

We would like to thank the editorial team for lending their expertise to make the book truly unique. They have played a crucial role in the development of this book. Without their invaluable contributions this book wouldn't have been possible. They have made vital efforts to compile up to date information on the varied aspects of this subject to make this book a valuable addition to the collection of many professionals and students.

This book was conceptualized with the vision of imparting up-to-date and integrated information in this field. To ensure the same, a matchless editorial board was set up. Every individual on the board went through rigorous rounds of assessment to prove their worth. After which they invested a large part of their time researching and compiling the most relevant data for our readers.

The editorial board has been involved in producing this book since its inception. They have spent rigorous hours researching and exploring the diverse topics which have resulted in the successful publishing of this book. They have passed on their knowledge of decades through this book. To expedite this challenging task, the publisher supported the team at every step. A small team of assistant editors was also appointed to further simplify the editing procedure and attain best results for the readers.

Apart from the editorial board, the designing team has also invested a significant amount of their time in understanding the subject and creating the most relevant covers. They scrutinized every image to scout for the most suitable representation of the subject and create an appropriate cover for the book.

The publishing team has been an ardent support to the editorial, designing and production team. Their endless efforts to recruit the best for this project, has resulted in the accomplishment of this book. They are a veteran in the field of academics and their pool of knowledge is as vast as their experience in printing. Their expertise and guidance has proved useful at every step. Their uncompromising quality standards have made this book an exceptional effort. Their encouragement from time to time has been an inspiration for everyone.

The publisher and the editorial board hope that this book will prove to be a valuable piece of knowledge for students, practitioners and scholars across the globe.

Index

CPSIA information can be obtained
at www.ICGtesting.com
Printed in the USA
LVHW062015100122
708212LV00002B/241

9 781647 402600